ANIMAL LANGUAGES

ANIMAL
LANGUAGES

THE SECRET CONVERSATIONS
OF THE LIVING WORLD

EVA MEIJER

Translated by Laura Watkinson

JOHN MURRAY

First published as *Dierentalen* in the Netherlands in 2016 by ISVW
First published in Great Britain in 2019 by John Murray (Publishers)
An Hachette UK company

1

A CIP catalogue record for this title is available from the British Library

This publication has been made possible with financial support from the
Dutch Foundation of Literature

N ederlands
N letterenfonds
dutch foundation
for literature

Hardback ISBN 978-1-473-67767-8
eBook ISBN 978-1-473-67769-2

Typeset in Sabon LT Std by Palimpsest Book Production Limited,
Falkirk, Stirlingshire
Printed and bound in Great Britain by Clays Ltd, Elcograf S.p.A.

John Murray policy is to use papers that are natural, renewable and
recyclable products and made from wood grown in sustainable forests.
The logging and manufacturing processes are expected to conform to the
environmental regulations of the country of origin.

John Murray (Publishers)
Carmelite House
50 Victoria Embankment
London EC4Y ODZ

www.johnmurray.co.uk

For Batyr, Nim, Peter and the others

CONTENTS

PREFACE

If you are lucky, you might meet an animal that wants to talk to you. If you are even luckier, you might meet an animal that takes the time and effort to get to know you. In my experience, most animals are quite willing to have a chat. They are also generous in what they are prepared to tell you.

You can enter into close relationships with some animals. Such relationships can teach us not only a lot about the animal in question, but also about language and about ourselves. Other animals have their own perspectives on life, and being able to see things through their eyes makes us see the world differently. Many people expand their horizons and have new experiences by travelling and getting to know different cultures, but there are many cultures waiting to be found on every corner – from ants and pigeons and cats to hares and cows.

The origins of this book are in my childhood, when not only people but also cats, guinea pigs and horses played a major part. In particular, Joy the pony – with whom I shared my life between the ages of eleven and sixteen – made me realise that it is possible for humans and other animals to have an extensive shared language. In my early adulthood, Pika the dog taught me about the language of canines and about what matters in life. Without Pika, this book would not exist. At the moment I live with a dog and a cat, Olli and Putih, who help me to think and to play.

When I was studying philosophy, I was surprised by the almost complete absence of animals in the Western philosophical tradition. Thinking has long been perceived as an activity *for* human beings and *about* human beings. But this is changing; animals are increasingly being considered, particularly in ethics, and more recently in political philosophy. Language, however, is still largely unexplored territory: the philosophy of language has devoted hardly any attention to animals. This is unfortunate as language can give us insight into animals, and non-human animals can give us insight into language. Research into animal languages helps us to see other animals, and ourselves, in a different way.

INTRODUCTION

Alex the grey parrot knew more than a hundred words. He used these words to demonstrate that he could, for example, count objects and separate them into categories. Alex also made jokes and employed words to influence the behaviour of the people around him.[1] Chaser the border collie has learned the names of more than a thousand toys and understands grammar. Dolphins living in the wild call one another by name. Prairie dogs have an extensive language for describing intruders, which they use to describe the size of humans, the colour of their clothes, any objects they have with them and their hair colour. Elephants in captivity can speak in human words. Wild elephants have a word for 'human being', which indicates danger. The languages of whales, octopuses, bees and many birds have a grammar. The mantis shrimp communicates using colours and has twelve colour channels, while humans have only three.[2] Dogs, unlike their wild cousin, the wolf, can understand human gestures and read emotions on human faces.[3] Marmosets take turns in conversation and teach the same skill to their offspring.[4]

Humans have devoted attention to animal language and communication ever since the ancient Greeks, but ethology, the scientific study of animal behaviour and communication, really took off around 1950, and in recent years there has been increased focus on animal language. The latest research shows that other animals communicate with one another in a much more complex manner than was previously thought. But despite this, little has been written about the significance of this discovery for animals and for our understanding of language. Can we call the communications of other animals *language*? Can we speak with other animals – and, if so, how? Is human language special, or are all languages special? What even is language?

My aim in this book is not to provide an overview of all animal languages – we still know very little about many types and there is a huge number of different species, each with its own language or languages. But here I explore empirical research into animal languages and the philosophical questions it raises. My aim is to show the wealth of animal languages all around us, and to explore how learning about them can change the way we think about animals.

Animal intelligence has long been measured in terms of human intelligence. Experiments have, for example, investigated how good animals are at solving puzzles in comparison to humans. Animals will never score as well as humans on these kinds of tests because their senses have developed differently – they need other skills to survive. But the reverse also applies: in ant terms, humans are probably not very bright, as they are not as good at working together; in pigeon terms, humans have poor spatial awareness; in dog terms, humans are unable to navigate by scent. In Chapter 1, I look at experiments that have attempted to teach animals to speak in human language and explore what they reveal about how language works.

In biology, intelligence is now understood as the ability to deal with species-specific challenges.[5] Animal communication is geared to their specific living environments and based on their physical and cognitive capacities. Whales, for example, often use sound as it travels quickly underwater; scent and sight are less useful in the ocean. By using very low-pitched sounds, elephants can maintain contact over a number of kilometres. Bats, on the other hand, use very high-pitched sounds to read their environments when navigating and hunting. These creatures have also developed very complex communication systems, which in some ways are similar to human language. In Chapter 2, I encounter animal languages in the living world and explore them in greater depth.

As animals do not usually express themselves in human language, humans sometimes believe we have no way of knowing what they are thinking. We can understand people because they speak; language gives us insight into their inner world. Animals cannot speak, so they will always remain a mystery. But we might also wonder whether we humans ever truly understand what other humans are thinking or feeling. Language can be misleading: someone can say that they love you and then later deny it. Misunderstandings can arise: someone might say that they love you, and you understand it romantically, but they meant just as friends. Language is not unambiguous, and neither are people. We can never have hard evidence of what people are thinking. In fact, some philosophers say, we can never have proof that they are thinking at all. Furthermore, we might ask why belonging to a particular species should determine our understanding of someone else. Humans like to categorise, but although other animals express themselves in different ways and perceive the world differently, there is still enough that we share. Species does not determine under-standing; social factors are as important. If you know an animal well – for example an animal companion with whom you share a household – you can often understand him or her better than you can under-stand a human from a completely different culture. In Chapter 3, I discuss conversations between humans and the domesticated animals that we share our lives with – our dogs, cats, guinea pigs and parrots, and farmed animals such as sheep, pigs and cows. Then in Chapter 4, I look at the role of the body in thinking, and I develop a phenomen-ological approach to animal research.

In Chapter 5, I explore the structure of animal languages in greater depth. It was long thought that only human languages had grammar and that animal languages were primarily a direct expression of their emotions. Recent research, however, has shown that this is not the case; animal languages sometimes also have complex structures, can be symbolic and abstract, and can refer to situations in the past or the future, or beyond the reach of animals in some other way.

One of the ways in which animals communicate with one another is by playing, and when playing, animals can also say something about their play. This is what we call metacommunication: commu-nication about communication. In Chapter 6, I look at the relationship between play, language, meta-communication and rules, and I discuss the morality of animals.

It may seem far-fetched to think about the language of animals – as if there were a huge gap between our forms of communication and theirs, and as if human language were more elevated, something that animals will never be able to achieve. However, not that long ago, women were thought to be irrational and incapable of making political decisions.[6] Colonised, non-Western populations were once not taken seriously as participants in discussions either. The property rights of Aboriginals in Australia were not recognised, for example, because this was not in line with the European settlers' system of laws and regulations. In Chapter 7, the conclusion, I discuss the role of language in politics. Thinking about animal language and the use of language with animals may help us to form new communities and relationships, and to look critically at the position of animals in our society.

When you write in language about language, or think in language about language, that language always influences you, which makes the study of language complicated. Wittgenstein compares it to repairing a spider's web with your fingers.[7] Language can mislead us because the form of the language equates things that are not the same. Take, for example, the word 'animals'. This makes it appear as if a dividing line exists, with all humans on one side and all animals on the other. But as the philosopher Derrida argued, a gorilla and a spider have less in common than a human and a gorilla.[8] The ancient Egyptians had no collective noun for all animals other than human beings, although they did have names for different species.[9] As a result of the fact that we have a word that captures all animals, we experience the line between humans and other species of animals more strongly. This perception, in turn, has the effect of reinforcing anthropocentrism, the notion that humans are the centre of existence,[10] and this can lead to oppression or even to violence against animals.

Words have power. The words that we use reflect – and influence – beliefs that exist in our culture. Language both expresses and shapes reality. To indicate that there is a line of continuity between humans and animals, animal philosophers often write about humans and 'other animals' or about human and non-human animals.[11] In this book, I use both terms; if I use the word 'animal' for other animals, I do so because it is concise and people know what it refers to. I am not saying that humans are not animals, or that humans are special as a species. All species are special in their own way.

Language not only misleads though; it can also build a bridge between different worlds. If we learn more about animals, perhaps we will be able to interact with them more successfully. Some humans will want to treat them better. As we understand ourselves and the world by using language, thinking about language is a promising tool in interaction with other animals. By gaining a better understanding of what they are saying, by looking at them and listening to them better, we can gain more insight into their worlds and experiences. By explaining better what we say – in ways that animals can understand – we can form new shared worlds. This will not result in all animals and all humans living together in complete harmony, just as humans are unable to live in harmony with all other humans. However, it could help us to find solutions for certain practical problems associated with living together – and such coexistence is unavoidable – and to seek out new relationships in a world dominated by humans.

Humans who write about animal language are often accused of anthropomorphism, the attribution of human traits to other animals. This is considered an unscientific and undesirable approach that projects a human view onto an animal.

Although such anthropomorphism does, of course, occur, this does not mean that we can never say anything about the thoughts or emotions of other animals or that we are automatically humanising them when we study particular characteristics. Existing concepts can actually help us to research other animals, as long as we remain critical and open. Besides, a certain degree of anthropomorphism is inevitable. As humans, we naturally have a human view of things. We have no access to an objective reality, a point in space from which we can see everything. Denying human characters in other animals – also known as 'anthropodenial' – can just as easily obscure the picture.[12] For a long time people questioned whether animals – and babies – could suffer pain. Few scientists deny this now, but this scepticism has resulted in suffering for many animals.

Language, philosophy and the world

Language plays an important role in how we think about people. Many philosophers in the Western tradition consider human language to be unique, and some even believe that language is fundamental to what makes us human. For Aristotle, command of language was necessary to make a distinction between good and bad, and so it determined who could belong to the political community.[13] Descartes believed that we can deduce from the fact that animals are unable to speak that they do not think.[14] The Enlightenment philosopher Kant concluded that animals had no *logos*, or reason, and therefore fell outside the moral community.[15] For the phenomenologist Heidegger, language was so important for our place in the world that those who have no language cannot die; they simply disappear.[16] All these philosophers defined language as human language, automatically excluding other animals. For them, language was connected to thinking itself, and viewed as an expression of reason.

These are still important questions in modern human society and politics. As non-human animals do not speak in human language, humans believe them incapable of acting politically, and this has consequences for their position in political and legal systems. If we do not understand animals, it is often assumed that their communication is not meaningful, and when they do not understand us they are thought to be stupid. It may seem logical for animals to have no rights and not to be heard by humans; human society prioritises the wants and needs of humans. The problem is that humans determine the lives of many animals to a great extent. Domesticated animals live with people and often have little freedom to make choices or to develop, while wild animals deal with human influence, with human populations occupying or polluting their territory.

The way we think about animals is connected to how we treat them. Take the example of Descartes, who thought that animals have no soul. He deduced this from the fact that they have no intellect, which in turn he deduced from their inability to talk.[17] Even deaf people who are unable to communicate with their voices, he writes, can still express themselves in human language in one way or another. According to him, animals are truly dumb, in both senses of the word, because they are completely incapable of expressing themselves in this way. Animals who repeat words – he gives the example of the magpie – do so on the basis of urges that motivate them to perform a certain action for a reward. Descartes believed that the body was purely mechanical, something that works like a clock: since animals have no soul and are therefore just a body, they are, in fact, a kind of machine. For that reason, he called them *bêtes-machines*. And because animals are only bodies, they can suffer no pain. They might scream when someone puts a knife in them, but this is not an expression of pain, purely a mechanical reaction. Descartes was interested in how bodies work and was an advocate of vivisection, which placed him at the start of the animal experimentation that still goes on today.

Determining whether other animals have languages or not might seem mainly a matter of doing empirical research. However, information from these studies always has to be interpreted. Research questions determine the answers other animals can give, and social bias colours these questions.

Philosophy can be a tool for investigating how things really work. On the one hand, this is a critical project: existing judgements and opinions are not automatically true just because many people believe them. On the other hand, it is experimental: thinking can place experiences in a new light, changing our understanding of the world. Wittgenstein states that the task of philosophy is to make us look differently at reality. Thinking about language and animals can help us to look differently at animals and language.

In this book I will make use of various types of insight: empirical research in biology and ethology, understanding gained from new academic disciplines that are focused on animals, such as animal studies and animal geography, and different branches of philosophy. My starting point is that animals have language. This is contrary to what was believed for a long time and it underpins the theoretical standpoints that I employ. I will deal with positions that are critical of current thinking about humans and animals, reinterpret positions from the Western philosophical tradition in relation to animals and discuss other literature based on the notion that communication with animals is possible, and that animal languages are worthy of study. The fact that animals express themselves differently from us does not mean that their utterances are not meaningful. Refusing, on principle, to take them seriously because they belong to another species is a form of discrimination, an expression of speciesism. Dolphins, for example, are social animals and are known to communicate frequently with one another. Their language is hard for us to understand and new technology is being employed as a tool to record and interpret their high frequencies. Who knows if we will ever be able to understand exactly what they are saying? But it would be unscientific, not to mention arrogant, to decide in advance that their communication is less meaningful or complex than our own.

Investigating language demands that we examine prevailing prejudices and, where necessary, adapt them. The questions that are asked determine the answers that animals can give. If you assume that animals have no language and cannot communicate meaningfully, the research you do will most probably prove exactly that. But if you assume that animals do communicate, and maybe in complex ways, you will ask different questions. Researching language is not only important for finding out how animals communicate with one another, but also for investigating how they communicate with us. Concepts and ideas that have been developed in philosophy can serve as tools for clarifying existing communication and for inspiring others to think more deeply for themselves.

CHAPTER

SPEAKING IN
HUMAN LANGUAGE

It is well known that parrots are capable of learning a large number of words. They often repeat human words, which is why we call humans 'parrots' when they copy someone else's words. A few years ago, on the back page of Dutch newspaper *NRC Handelsblad* there was an anecdote from a vet about a parrot with a bad cough. The vet examined the parrot and found nothing wrong, but the parrot stayed in overnight for observation. When the parrot's human came to pick him up, he stopped outside first, to smoke a cigarette. The parrot was perfectly imitating the resulting smoker's cough.[1]

Because of the way their bodies are built, parrots are one of the few species capable of repeating human words, but it had been assumed that their linguistic capability was no more than a capacity for imitation. You could teach a parrot to say 'hello' as a greeting, but that was about as far as it went. In 1978, the psychologist Irene Pepperberg began an experiment in which she worked with Alex, an

African grey parrot.[2] She wanted to investigate if parrots were capable of learning language, and she based her assumptions on how communication works among birds. Language learning in parrots is strongly associated with actions. Pepperberg taught Alex words by letting him decide his rewards for himself, and always connected those words to his use of them. So, by learning words, Alex gained more control over his surroundings. He could indicate which snacks he wanted as a reward and when he wanted to take a break or go outside. Pepperberg made use of this to teach him new words, which gave her insight into Alex's thinking.

In this way, Alex developed a vocabulary of around 150 words and was able to recognise fifty objects. He could understand and answer questions about the objects. He learned to recognise colours, shapes, materials and functions. For example, he knew what a key was for. He recognised new keys as keys, even when they were a different shape. He also showed understanding of concepts such as 'same', 'different', 'bigger', 'smaller', 'yes', 'no' and so on. When Alex was bored, he sometimes gave the wrong answer on purpose. Pepperberg once asked him what was the colour of block three, and he said 'five', as she had asked him this question before. He went on repeating this until she asked him for the colour of block five. 'None,' said Alex.

He could also count, understood the concept of 'zero', knew how sentence structure works and could combine words. When Pepperberg and her assistant made mistakes, Alex corrected them. He sometimes practised words when he was alone. Alex once asked Pepperberg what colour he was himself, a pretty existential question for a parrot.

Ornithologist Joanna Burger has described a different kind of relationship with a parrot.[3] She adopted thirty-year-old Tiko, who changed from a negative, grumpy and sometimes even hostile bird into a very loving parrot. Tiko saw Burger as his partner, courting her in the mating season and fighting her husband when he came too close. His previous humans had not taught him to speak and Burger did not attempt to either. However, there was plenty of communication, some of which was accompanied by words, and Tiko understood many of them. When Burger announced that she was going to work, which happened at different times of the day, he would go to his room. He also used the standard parrot vocabulary, such as 'hello' and 'good boy'. When he was not feeling jealous, Tiko liked to whistle a duet with Mike, Burger's husband. He did this when Mike played the guitar and also when he thought Mike was angry with him for breaking or stealing something. Tiko knew that whistling would distract Mike and put him in a good mood. He made use of various types of gibberish, which indicated different moods. When Burger was on the telephone, he liked to join in loudly.

Ethologist Konrad Lorenz has also written about parrots teaching themselves words.[4] He observed that in addition to their sensitivity to human habits and behaviour, which parrots express by saying 'good morning' at the right moment, for example, events that have an impact on them can also result in them spontaneously retaining a particular sound. Lorenz describes this in an anecdote about Papagallo, an Amazon parrot. Many birds are scared of things that come from above, because it reminds them of a bird of prey. When the chimney sweep came for the first time, Papagallo was terrified. The next time – a few months later – he called out 'the chimney sweep is coming, the chimney sweep is coming' when he saw the sweep approaching. He had probably heard the words 'chimney sweep' from the cook and remembered it because the previous visit had made such an impression on him.

Mimicry – imitation of another species – extends further in parrots than just the use of words. Burger describes parrots waving to people with their feet or acting as if they are putting on their coat when they leave. Others nod or shake their heads at the right point in a conversation. According to Burger, parrots in the wild do this too. Recordings of two wild grey parrots reveal more than 200 different patterns, twenty-three of which were an imitation of other birds, and one of a species of bat. This mimicry is a handy trick in the wild if you want to deceive other birds, either because you want to steal something from them or because you do not want to be attacked yourself.

In social psychology, mimicry also denotes the phenomenon of people unconsciously imitating one another. Humans spontaneously mimic body posture and actions such as smiling, yawning, crossing legs, putting a hand under the chin; all these things are contagious. Humans often mirror someone else unconsciously and stop as soon as this is pointed out to them.[5] Humans who have a sense of connection with each other or who are part of the same group engage in mimicry more frequently. Mimicry can also generate greater solidarity as humans who mimic understand each other better and their emotions become better attuned.[6] Mirror neurons have been found in monkeys: brain cells that show activity when the monkey makes a movement, but also when the monkey sees another monkey making the same movement. In humans, there is the same overlap in areas of the brain when observing someone else perform an action, performing an action and thinking about an action.[7]

Just as with humans, for Alex and Tiko mimicry can have different functions in different contexts. When faced with an enemy, it can be a form of self-defence. When mimicry occurs in a relationship with a human they know, as described by Burger, it could perhaps work in just the same way as between humans: in order to become better attuned or as an expression of closeness. Pepperberg and Burger have shown that parrots can develop language both among themselves and in relation to humans. That language is different from the language of humans, and yet meaning can be conveyed between a human and a parrot. Pepperberg is not arguing that Alex can speak English, simply that he uses words and concepts and therefore shows understanding and intelligence. She makes this distinction to show how the use of words by parrots is perhaps even more strongly connected to meaning than it is with humans. Pepperberg's systematic research undermines the idea – long popular among animal researchers and lay people – that parrots act only out of instinct.

As everyone knows, dogs and their humans can begin to resemble one another after a while. Mimicry is perhaps the explanation for this phenomenon: human and dog unconsciously imitate each other's facial expressions and body language and start to look similar, even though the form of their faces and bodies is very different.

Chimpanzee children

In the 1920s, humans became interested in researching language and its development with the help of non-human primates. Humans and other primates are genetically closely related and humans wondered if the ability to speak was primarily a question of nature or of culture. In an attempt to find out, a new kind of experiment was devised, in which chimpanzees were taken into human homes and brought up as human children, usually by a married couple of animal researchers. The first chimpanzee to be raised in this way was Gua. She moved in with Luella and Winthrop Kellogg in 1930 when she was seven and a half months old and was brought up in the same way as their son, Donald, who was ten months old at the time. Gua did not learn to speak.[8] Viki, who was taken in by Keith and Catherine Hayes in 1944, learned – partly as a result of intensive speech therapy, which involved manipulating her lower jaw – to say four words.[9] As these two experiments had little success, it was thought at first that other primates were not intelligent enough to learn language. Later it was believed that they were unable to pronounce the words clearly because chimpanzees have a different larynx from humans. So new experiments focused instead on sign language.

Washoe is the most famous chimpanzee to grow up with humans. She was born in the wild and taken from her parents by the US Air Force for space experiments. Allen and Beatrix Gardner took her into their home for an experiment run by the University of Nevada. They brought her up like a child: they dressed her in clothes and she ate at the same table as them, went for drives with them in the car, played outside. She had toys, books and her own toothbrush. Learning sign language proved a success. Washoe not only learned what she was explicitly taught, but also watched and acquired gestures that humans made to one another, and she came up with words herself (for example, combining the gestures for water and bird to make swan). She understood that the sign for 'dog' could refer to all dogs, and she could form simple sentences.[10] When Washoe was five, the Gardners decided that they had had enough, and she was moved to a research institute and lived in a laboratory until her death.

At the laboratory, further research was done into

her capacity for language; in total, she learned around 250 signs. The researchers also learned about what she thought and felt. Washoe recognised herself in the mirror and was startled when she met other chimpanzees. When new students came to work with her, she deliberately signed more slowly so that they could keep up more easily. One of Washoe's keepers was pregnant and did not come in for a few weeks. When she returned, Washoe was upset with her and ignored her. The keeper decided to tell Washoe what had happened and signed to say that her baby was dead. Washoe first looked away and then at her, carefully making the sign for crying. Chimpanzees do not cry, but Washoe had learned that it was what humans do when they are sad. The keeper said later that this simple gesture told her more about Washoe's inner world than the artificial sentences that she could form.

Nim Chimpsky, another chimpanzee, was also raised by humans, but learned to use far fewer signs than Washoe.[11] Herbert Terrace, who wanted to invalidate the research into Washoe's linguistic proficiency, led the experiment in which he was involved. He had Nim brought up in a foster family with seven human children, in an environment that gave him little structure. When Nim hit puberty, there were a number of incidents with him biting the people around him, and finally Terrace had him taken to a laboratory, where the experiments continued. Nim learned to use 125 signs. There was doubt, however, about whether these were an expression of linguistic proficiency as he had learned them through operant conditioning: it was argued that he signed because he received a reward when he did so correctly, not because of any understanding. Terrace argued that Nim did not understand what he was doing, which was, of course, the intent of his research. When the research was over, Nim was taken to a pharmaceutical laboratory, where he was used for testing. Eventually he was moved to a shelter, where he died at the age of twenty-six.

Other chimpanzees took part in research in non-household environments. In 1967, Sarah and three other chimpanzees began learning how to analyse and produce sequences of symbols in a laboratory.[12] With the help of a board with plastic symbols, they learned some grammar and simple sentences. Sarah was studied for twenty years and is one of the best-known research chimpanzees. Other famous chimps are Kermit, Darrell, Bobby, Sheba, Keeli, Ivy, Harper and Emma. A number of them now live at Chimp Haven, a shelter that aims to offer a good life to all the remaining laboratory chimps in the United States.

Koko and Kanzi

In addition to the research carried out on chimpanzees, the linguistic proficiency of other primates has also been studied. Koko the gorilla was born in 1971 at San Francisco Zoo. Francine Patterson used Koko for her doctoral thesis into language in gorillas and worked with her until she died.[13] Koko became famous because she had a cat as a companion (she called the cat 'all ball' because it had no tail). She knew more than a thousand signs of Gorilla Sign Language, as Patterson calls the language she taught her (GSL is like human sign language, but gorilla hands are shaped differently, so some signs vary), and understood more than two thousand spoken human words. Koko liked to make jokes and had a good memory. The signs she learned enabled her to convey her memories, which gave humans insight into the way gorillas experience the world. For a while she lived with Michael, a male gorilla who knew around 600 signs, some of which he had learned from Koko. Michael used the signs not only to describe objects, but also to communicate his emotions, dreams and memories, and to tell lies. One of the memories that he conveyed with signs was the murder of his mother by poachers in Cameroon when he was still very young.[14] Michael also loved to paint.[15]

Kanzi the bonobo learned signs by watching videos of Koko. His trainer only realised he could do this when he suddenly started using sign language to communicate with an anthropologist. Kanzi showed that bonobos can learn language not only from humans or bonobos, but also by looking at other primates. Kanzi had already learned to use lexigrams – a symbol on a keyboard used in the artificial primate language Yerkish, which is primarily employed in communication with chimpanzees and bonobos – by watching his adoptive mother Matata's lessons. Kanzi knows 210 lexigrams, and when he hears a word spoken on the headphones, he presses the right button. Kanzi likes to make omelettes, can play Pacman and is a talented toolmaker – he can make good, sharp knives out of stones, for example.[16]

When Kanzi uses the lexigrams, he makes sounds. Although bonobos are unable to speak words, it seems as if Kanzi is trying to do just that. With Kanzi and Koko, humans also wonder, as they did with Washoe and Nim, if they are actually using language or simply repeating words. While Pepperberg interacted with the parrot to see how the creation of meaning was possible, this animal research focused strongly on teaching human language. Patterson is convinced that she and Koko understood each other and says that the gorilla herself understands the signs she is making. If you watch a video of Patterson and Koko, you can see that this gorilla and human are well attuned to each other.

The philosopher and animal trainer Vicki Hearne writes that mutual understanding between humans and other animals is possible in working relationships.[17] Dogs, for example, experience the world in a different way from humans – scent is important to them, while we rely more on sight – but words and gestures gain meaning when humans work with dogs, and communication and understanding become possible. Humans and other primates are more similar from a physiological point of view than humans and dogs, so it seems quite conceivable that communication and understanding can take place between them. However, the role that human words can and should play in this communication is debatable. Kanzi the bonobo knows a large number of words and can use an artificial language to make some of his wishes known to the people around him. The humans with whom he communicates use the same artificial language. Hearne points out in her accounts of communication with dogs and horses that gestures, body posture, eye contact, touch and other physical forms of interaction are more important in communicating with other animals than the use of human words. The context also matters: an intelligent and sensitive animal in a small cage in a laboratory, without others of its kind, will most likely react differently from an animal in a normal social environment, and this artificial setting also influences how the human in question responds. Words, gestures and other forms of communication gain meaning in the social context within which they are used. When we think about the linguistic proficiency of primates, we need to look both at their answers to the questions that are asked and at the questions themselves.

The research into Washoe, Nim, Sarah and the others was in the first instance about the origins of human language and its presence in other primates. The idea behind this research is that humans are a species of super-evolved primates – the pinnacle of creation – and that other primates provide insight into our history. This is not correct from an evolutionary point of view, however; humans and other primates descend from a common ancestor, and humans do not descend from other living primates. The picture that is painted of other primates is also problematic. They are not failed humans; they are beings with their own abilities. Humans and other primates are similar in many respects, and different in others. If we want to know what those similarities and differences are, we need to develop research based on their view of the world.

Scientists now believe is not true that non-human primates are unable to pronounce words because of the shape of their larynxes.[18] We do not know exactly why they do not speak, but a tiny area of the brain has been pinpointed that is connected to this ability, and it appears to be genetically determined. It is equally untrue that other primates are incapable of complex communication because they cannot speak in human words. As humans do, chimpanzees use countless gestures and vocalisations among themselves. By 2015, sixty-six chimpanzee vocalisations and eighty-eight gestures had been mapped. Researchers have used this information to compile a dictionary.[19] Tapping another chimp, for example, means 'stop that', flinging a hand to one side means 'move away' and lifting up an arm means 'give me that'. Nibbling on a leaf is an invitation to flirt. A big hug is an invitation to go somewhere, as is scratching loudly. Banging one object against another is an invitation to come closer. Different gestures can have different meanings, but it is also possible that there are subtle distinctions that humans do not notice. Chimpanzees can also use gestures to show people the way to food.[20]

Trends sometimes develop within groups of

chimpanzees, such as wearing a blade of grass in your ear, which was all the rage with chimpanzees at a Zimbabwean reserve. Primatologists from the Max Planck Institute for Psycholinguistics in Nijmegen, the Netherlands, have been researching this phenomenon since 2010. Julie the chimpanzee started wearing a blade of grass in her ear back in 2007. Other chimpanzees began to imitate Julie, particularly if they spent a lot of time around her. The wearing of a blade of grass is the first known example of fashion among chimpanzees as it has no apparent purpose and is purely decorative. When Julie died in 2013, wearing blades of grass became less popular, although some chimpanzees continued to do so.[21] Chimpanzees have other traditions too, such as methods for using sticks to catch termites. They also make tools out of stones, which means they have entered their Stone Age.

Dolphins and whales

At the beginning of the 1960s, neuroscientist John Lilly set up a laboratory on the Caribbean island of St Thomas to study the language of dolphins.[22] Dolphins can make human-like sounds through their blowholes, and Margaret Lovatt, a young woman with an interest in dolphins but with no scientific background, wanted to research if, in a close relationship and with a lot of training, they could learn to speak. So in 1965 she moved to an underwater house, together with the young dolphin Peter, one of three dolphins kept at the aquarium. She gave him speaking lessons twice a day. Peter worked hard. He found her name difficult to say, for example, and tried to blow the 'M' underwater with air bubbles. However, Margaret soon discovered that it was not the speaking lessons that gave her most insight into Peter's thinking. She learned more when they were just hanging around. Peter was very interested in her anatomy, for instance. He stared at her limbs for a long time and seemed to be trying to understand how they worked.

The study lasted six months. During that time, Lilly began to experiment with LSD on the dolphins. As a result, and because of the publicity surrounding sexual activity between Peter and Lovatt, the project lost its funding. As an adolescent male, Peter often became sexually aroused, which got in the way of the training. At first, Lovatt sent him to the other aquarium with a female dolphin (he had to be hoisted up there in a kind of lift). After a while, though, she began to use her hand on him instead; it was faster and she did not object to doing it. The story spread, though, resulting in an article in *Hustler*. According to Lovatt, the reports were misleading, but the damage had been done. The dolphins were moved to a smaller laboratory, without daylight, and without Margaret. A few weeks later, Lovatt received a phone call from Lilly. He told her that Peter had committed suicide. Dolphins, unlike humans, breathe intentionally. They have to go up to the surface whenever they want to breathe. When life becomes unbearable, they take one last breath and sink to the bottom, where they remain.[23] Lilly continued to research communication with dolphins, in scientific ways such as with music, but also using more mystical methods such as telepathy. Through his contact with dolphins he learned that captivity is harmful to them, and he later became an advocate of animal rights.[24]

More has since been learned about the language of dolphins, which is thought to be very complex. We do not know exactly how complex as we cannot hear many of the sounds, which are outside of our auditory range and equipment for recording them has not been around for long. Researcher Denise Herzing makes use of digital techniques to translate dolphin language into human language, and vice versa.[25] In 2013, she succeeded in translating a word for the first time: 'sargassum', a kind of seaweed, using a dolphin translation device. Earlier research into dolphin language worked in the same way as the primate research I previously discussed: dolphins learned symbols and the meanings of words. They learned that words in a sentence mean something different when you change their order, and they learned to understand gestures and body posture. The dolphin translation device gives humans the opportunity to communicate more extensively with dolphins and runs parallel to other research into dolphin behaviour. In order to interpret signals properly, we need to understand when they are used and how they fit within the wider setting of the dolphins' lives. Different groups of dolphins have their own dialect, or even language – an indication that language is culturally transmitted and does not purely stem from instinct or physicality – so it will be a long time before we can truly communicate with them. Much remains to be discovered; only time will tell the scope of their interaction.

Noc, a beluga whale, was captured in the late 1970s for the Navy Marine Mammal Program, which still exists. Whales and dolphins are used, for example, to detect bombs underwater with their sonar, and Noc was employed to search for torpedoes in the Arctic. These whales and dolphins are trained using voice and hand gestures. One day, Noc's trainer heard people chatting underwater. There was no one around. Later it happened again.

It turned out to be Noc, who was mimicking humans.[26] He spent his life in captivity, and his trainer thought this was a way to form a stronger bond with the people around him. After four years, Noc stopped talking. He died of meningitis at the age of twenty-three.

Elephants

The Asian elephant Batyr and the Indian elephant Kosik, both of whom were kept in zoos, went a step further than Noc, actually saying human words.

Batyr was born in 1969 and lived all his life at Karaganda Zoo in Kazakhstan without ever having seen another of his species. Batyr spoke for the first time just before New Year's Day in 1977 and went on to develop a vocabulary of more than twenty sentences. For example, he would say 'Batyr is good' and use words like 'give' and 'drink'. He also used the words 'yes' and 'no', and knew several swear words. He would change the sound of his name depending on his mood and use his trunk to alter the position of his tongue. At night he spoke quietly to himself in his cage, without using his tongue but making unarticulated sounds instead. It was not only the sounds of humans that he mimicked, but also those of dogs and mice, and mechanical noises.[27]

Kosik lives in an amusement park in South Korea and has taught himself a number of words, including 'hello', 'sit', 'lie down', 'no' and 'good'. Koreans who hear the recordings can clearly understand what he is saying. Scientists are not sure if he understands what he is saying: he knows what the word 'sit' means but when he says it he does not expect the zookeepers to sit, so he is not using it as a command. Between the ages of five and twelve – a crucial period in an elephant's development – he was the only elephant in the park, and scientists believe that he started mimicking people's speech in order to form a stronger bond with them. Like Batyr, he uses his trunk to speak, and the sounds are at the same frequency as his keeper's voice. He is now living with a female elephant. He speaks to her in elephant language, and continues to use human language with the people around him.[28]

While some sounds made by dolphins are too high for human hearing, some elephant sounds are too low. Like dolphins, elephants live in complex social relationships and sound plays an important role in their communication. Elephants have two voices: they can speak with their mouths and with their trunks. Low-frequency sounds, also known as infrasound as it is below the threshold of human hearing, can travel longer distances than high-frequency sounds.[29] The sounds are audible up to four kilometres away, with loud calls being heard even up to as far as seven kilometres away. The discovery of these sounds solved a number of puzzles for elephant researchers, such as how males manage to locate females over long distances during the mating season and how families who are kilometres apart are able to find the same locations. In order to hear the infrasound, researchers play recordings at about three times the speed. Researchers from the Elephant Listening Project[30] believe that elephants have an extensive language, in which they convey not only information but also emotions, intentions and physical characteristics. They have specific sounds for acquaintances (they can distinguish between hundreds of individuals on the basis of that sound), and sounds, or words, for humans and for bees, for instance. Sounds are used to indicate family relationships. They probably also refer to abstract concepts.

One of the reasons why elephants can form such complex communities is that they have a good memory for events and for individuals. Female elephants live in groups, which the young males leave once they reach puberty. It was long assumed that males only had social contact with one another in the contest for land or females, but recent research has shown that they also form close friendships, and live in larger groups of friends.[31]

Relationships do not cease at death. When an elephant is dying, the other members of the group, often family, come and stand around the dying elephant, gently comforting him or her with their trunks. When the elephant has died, they sometimes try to hold him or her upright, or to push him or her back up. Then they cover the body with soil and leaves, and for years they return to places where elephants have died: elephant graveyards. They also show interest in the bones of strangers. The combination of their good memory and their involvement with dead family members suggest that they have an abstract understanding of death. Maybe further research into their language will throw more light on the subject.[32]

Research into the language and intelligence of elephants, and social relationships among wild elephants, could help us to gain better understanding of the talking elephants in zoos. Learning human words and using them in the right context should not be so incredibly difficult for elephants, given their intelligence. The fact that they do their best to imitate words correctly, in a way that is physiologically challenging, shows how important social contact is for them. Batyr, who never knew any other elephants and spent his entire life in a small space, must have been very lonely, and also very bored. The words he spoke in human language give us far less insight into his linguistic ability than the research of the Elephant Listening Project.

Calling each other

Ethologist Konrad Lorenz shared his life with a large number of animals, all of whom roamed free in and around the house.[33] He used cages only when his children were young – when he and his wife were unable to keep a close eye on them, they would lock up a child in the pushchair rather than the animals. Lorenz often raised birds himself; he became well known for a theory about imprinting. Young birds of some species consider whatever they first set eyes on when they come out of the egg – or in the days immediately following – as their parent, whether it is a human being or the real parent. Not all birds will immediately follow a human though. For birds like ducks, geese and swans, the sound made by the future parent is crucial. In order to rear these birds properly, Lorenz had to imitate the call of the mother. So Lorenz learned to speak duck.

As well as the call of the mother, call-notes play an important role in bird interaction and are seen as an expression of instinct, innate sounds made in certain given circumstances. Lorenz describes how instinct and intelligence are intertwined in a large number of species. Reacting to call-notes is innate for many animals: they do it automatically and do not need to learn anything in order to do it – just as a human child normally does not need to learn how to cry. At the same time, call-notes can have a cultural function. Calls are passed on to members of a group and creative birds can give them their own spin. Roah the raven was raised by Lorenz and, even when he was an adult and had found his own place to live with other ravens, he still often came by to go walking or skiing with Lorenz. As Roah got older, he became more nervous; he did not like to go back to places where he had once had an unpleasant experience, and he was scared of strangers. At such times he would fly low over Lorenz's head and call him as he would another of his kind, in order to warn him. The sound he used was not the call-note he made for other ravens though. It was Roah's own name, with human intonation – the sound that Lorenz used to call him. Lorenz writes that he could not have taught Roah how to make this human sound. Roah invented this call for Lorenz, which Lorenz sees as the only case he knows of an animal showing this sort of linguistic insight.

Ravens and other corvids have an extensive arsenal of sounds, which mean something different depending on intonation, pitch and speed, and can be used to indicate a variety of things, such as a distinction between individuals. A study by Michael Westerfield, a crow researcher, has shown that not only do crows have different sounds that mean 'human', 'cat' and 'dog', but they can also distinguish between two different cats. An old cat who does not hunt is indicated with a different sound from a young cat who might be targeting young crows.[34] Young crows will babble away before they learn to make real crow sounds; researchers compare this to the baby talk of human children.[35] Crows talk primarily to family members, but before and while eating there is plenty of communication with strangers, particularly when they experience difficulty in obtaining food. In one study, crows were confronted with beetle larvae in deep holes in a tree trunk. Strangers immediately entered into conversation with one another, probably to exchange information about the best technique for getting the beetle larvae out. Researcher Christian Rutz says that providing hard-to-reach food has the same effect on the group of crows he works with as putting a coffee machine in an office.[36]

Observing corvids has revealed that they are able to communicate in ways that humans once believed were unique to primates and cetaceans. They never forget a face – if they are angry with you, perhaps because you threatened their young, they will attack you whenever you go past.[37] Corvids hide their food, which demonstrates that they have good memories.[38] Ravens communicate not only with their voices, but also make use of gestures, to convey information about objects, for example.[39] Corvids can solve complex puzzles too. Researcher Alex Taylor demonstrated how a crow nicknamed 007 solved a puzzle in eight steps in order to get a tasty snack. First he used a short stick to get a stone out of a barred box, and then a second stone from another such box. He dropped the stones into a plastic container, and extracted a third stone, again from a box with bars, which he also dropped into the container. Their weight made the hatch open and then he could pick up a long stick to extract a piece of meat.[40] Crows (like ravens, magpies and some other birds) hold funerals when a member of their group dies, which involves them gathering, sometimes in large groups, around their dead friend or relative and making noise.[41]

From an established truth to language games

There are many ways in which humans have linguistic interactions with other animals. These practices are not exactly comparable to a natural language like Dutch or English, and yet they can certainly be interpreted as expressions of language.

Since Plato, the philosophical tradition has been in search of truth. The image of truth that he painted is universal and unambiguous. According to Plato, the truth was not to be found in the everyday, but in eternal ideas that could be perceived only with the intellect, reflections of which we see in the reality around us. This image of truth is accompanied by the idea that language is an unambiguous and pure reflection of that to which it refers and the notion that the concept of 'language' can be clearly defined and known. In this conception, 'language' has a precisely defined meaning that can be universally applied.

In his later work, the linguistic philosopher Wittgenstein rejected the idea that words have an unambiguous meaning and that language can be defined in one way.[42] According to him, it is not possible to give a definition of language, and such thinking also clouds how language and meaning work. Language is used in countless different ways, and the meaning of words and concepts, and of the word 'language', can differ depending on the situation.

In order to understand what language is, we have to study how language works, which we can do by studying the practices within which it is used. Wittgenstein makes a comparison with the word 'game'. There are many different games, which do not share a common characteristic that would allow us to define them. Some games have certain common characteristics, while others have different ones, yet when we play a game we know that it is a game. The concept of 'language' also consists of numerous ways in which language is used, but not all of these have one common characteristic that we can use to define it. So when Wittgenstein talks about 'language games', he does not mean that language is like a game or that people are always playing when they use language, but that the structure of the concept of 'language' is comparable to that of the concept of 'game'.

Wittgenstein's concept of 'language games' – which refers to the entirety of language, individual language practices and very primitive artificial languages – is appropriate for thinking about communication with animals as it does not give a fixed definition and is therefore suitable for studying a variety of linguistic actions. Language games extend beyond words alone to gestures, posture, movement and sound. Wittgenstein gives the examples of singing, praying, whistling a tune, making a joke and solving a sum. A serious sentence can turn into a joke depending on facial expressions, intonation and gestures. Someone who has not yet entirely mastered a language can still make themselves understood using hand gestures, for instance, even if the words that are used have the wrong meaning. And, according to Wittgenstein, meaning is closely connected to usage, very similar to what Pepperberg indicates in her communication with Alex. The situations I discussed above cannot be understood as a natural language, like Dutch, but certainly can be seen as language games between people and other animals.

Both the relationship between language and thinking and the relationship between language and reality are subjects of philosophical study. Many humans think that the ability to use language is located in the mind. But Wittgenstein shifts the focus from the mind to the relationship between language and the world, pointing out the role of social practices in particular. The meaning of an utterance does not come from outside (a higher power or a necessary structure of the world) or from the mind (conceived of as a closed space into which others cannot look). Language receives meaning through its use and so is always a public matter. Even if we think things to ourselves in words or write them down for ourselves, that social component is present – we have learned to speak and write from other people and the way in which we express ourselves is part of a tradition or culture; new twists are possible but something completely novel is incomprehensible. The emphasis on the relationship between usage and meaning provides a new angle from which to study language with and of animals, in which scepticism about other animals' thinking no longer plays a role. We do not need to know what is inside their heads to determine whether or not they speak; we need to look at how they use language, and take it from there.

It is also the case that our concepts have been formed partly in relation to other animals because we live together with other animals. Children learn what words mean partly through the deeds and actions of other animals, stories about those animals and interaction with them. The Australian philosopher Raimond Gaita has written a book about the animals in his life, [43] in which he discusses the influence of other animals on how we think about language. Language is essentially a social phenomenon and because many humans live in shared communities with humans and other animals, those animals play a part in the words we use. When we think about the meaning of a concept and the relevance of that concept for other animals, we have to take this into account. When we wonder whether animals feel pain or have intentions, as defined by humans, we are thinking the wrong way round. We learn what pain is partly by observing and speaking about the pain of other animals, so their pain is already part of what 'pain' can mean. Animals do not have to satisfy a particular cognitive standard in order to gain access to a concept, as through their thoughts and deeds they are already part of it.

Wittgenstein's ideas about language can help us to think about language with animals. His method can also shine new light on the question of whether animals have language at all. The way in which language is defined as a whole can also be seen as a language game, one in which reasonable adults define a particular form of linguistic expression as the true or real language. This way of thinking has a history and did not appear out of nowhere; it is a social practice, influenced by power relations. We can study the history of concepts and research the influence of social relationships on how concepts change. Take, for example, the concept of 'rights'. In the Greek *polis*, the city state, only free men had the right to make political decisions. Slaves and women did not, and animals and children most certainly did not. Various movements – such as the civil rights movement and the women's movement – have ensured that democratic rights apply to most adult human beings in the West. New rights were granted and the meaning of 'rights' expanded from something for a select group to something for all humans, at least in theory. Animal rights are certainly not generally accepted, but if animals were to gain rights, then the concept would change once again.

According to Wittgenstein, in order to investigate the meaning of language, we should study existing language games, and we do so by studying the practices within which they take shape. So when non-human animals are taught to speak in human language we should also bear in mind that an artificial setting is involved, within which they learn an artificial language. The relationship with their trainer is of importance, as is any relationship they might have with others of their kind, the method used, and so on.

Researchers have developed the K9 sign language for dogs, and have also taught them to work with lexigrams. When one dog was introduced to a guinea pig, he pressed the symbol for 'food' rather than, as the researcher expected, the symbol for 'play'.[44] This gives us some indication of this dog's thoughts – in this case that another household companion is seen as a possible lunch, rather than a playmate – but is far from telling us everything about how well the dog can communicate or about the species-specific linguistic capability. It only tells us something about how good an animal is at this specific language game. By contrast, dogs communicate with very complex scent signals, but this skill is often not considered to be language. Humans who aim to teach other animals to speak in human language are engaging in a language game in which human language is seen as the only real one and is used as a criterion against which linguistic competence and intelligence are measured.

Learning to speak

At the time of writing, five species of mammals – humans, bats, elephants, seals and whales – are assumed to be capable of learning how to produce new sounds.[45] These animals can learn human words, and some can learn to speak the language of other animals, or attempt to do so. For example, orcas are known to imitate the sounds of dolphins and to use this skill to communicate with them.[46] Parrots mimic the sounds of other animals, both for self-defence and to hunt those animals, and there are also other birds that can learn to make new sounds. It also seems premature to rule out other mammals. Tilda the orang-utan lives in Cologne Zoo. She can whistle like people and makes a whole range of sounds that appear human. These are completely different from the noises that orang-utans make in the wild and clearly resemble human speech, particularly in the rhythm and the alternation between vowels and consonants.[47] There are also videos on the Internet of dogs and cats imitating human sounds, but the scientific relevance of this is not yet clear. Kosik the elephant can pronounce words. Nevertheless, it seems problematic to look only at words when considering imitation. As animals express themselves primarily through gestures or by body language, scent or other expressions, perhaps this is where we should be looking for imitation as a form of communication.

Most talking animals belong to very social species. They speak for various reasons: in captivity it might be in order to strengthen the bond with their human captors or because it is the only language they hear. Speaking can give animals control over their environment – for example, when Kanzi uses lexigrams to ask for a pizza – like using a tool. It can also be a form of play – as in the case of a captive Japanese beluga who appears to speak in order to involve humans in his game.[48] In the wild, animals use their learning ability partly to strengthen existing relationships and to impress or court others. There are also animals who imitate other sounds – the starlings at the railway station in Rotterdam mimic the departure signal of the Sprinter trains, for instance, and there are birds who make telephone sounds. Scientists believe they do this to impress others.[49]

Vocal mimicry is the basis of human language. Our ability to imitate allows us to learn and reproduce a large number of words and sounds and is the reason why our vocabulary is so extensive. At the same time, learning involves more than imitation; words cannot be understood independently of their context. Wittgenstein begins his *Philosophical Investigations* with a scene in which a child is learning a language by attaching words to objects. 'Table' refers to a table, and 'chair' to a chair. This is indeed a way in which language works, he writes, but it is not the only way. When you learn a language, it is not sufficient to learn only the words and the related objects or actions because words receive meaning through practices. The meaning of a word can differ depending on the situation, and it means that in order to use language well we need to know how a word is used. This extends beyond mere imitation, in humans and in animals.

When the context – or the language game, to stay with Wittgenstein's terminology for a moment – involves learning words in exchange for rewards, the animal's ability to do so says little about linguistic proficiency, but more about their skill in learning words in exchange for rewards. Animals who are unable to pronounce any words are automatically excluded from such language games, and animals who are not good at mimicry because they do not need it in their daily lives are also at a disadvantage. But although learning human words is, therefore, usually an artificial process and devised by humans, it can still teach us things about animals: for example, about the way they learn, about their thinking, cultures, and about their memory. Michael the gorilla for example signed about memories of experiences he had in the wild as a very young gorilla, which indicates a narrative identity (an understanding of self over a longer period of time) and an episodic memory (one that records personal experiences, part of the long-term memory). Here it is also important to notice that the cases above involve more than vocal mimicry. The most meaningful communication in the cases of Koko and Washoe consisted of gestures and eye contact, the moments when the other animal made emotional contact with the human, and vice versa.

CHAPTER

CONVERSATIONS IN THE
LIVING WORLD

Humans walk along the high street talking to each other and into their phones, sending messages, flirting with passers-by, bumping into one another, swearing. Up above, a male pigeon on a window ledge coos at a female. Seagulls circle overhead, shrieking, on the lookout for chips. In a crack between the street and a building, ants walk along, leaving traces of scent to let other ants know that there is food nearby. Mice are living in the wall, singing at a pitch so high that others can barely hear it, if at all. In front of the butcher's a dog waits for a chunk of sausage. He makes eye contact with a passer-by. In a half-finished underground train tunnel, one rat uses pheromones to warn off another. In the nearby canal, a perch seeks contact with another perch by vibrating its swim bladder. On the water, young coots call for their mother. A duck begs for bread.

A city appears to be populated mainly by humans, but other animals are everywhere, communicating with members of their own and other species. We naturally understand some of these animal languages, but others are a complete mystery, and there are all kinds of degrees of understanding in between. There are expressions of language that we hear or see and do not understand, and there are utterances that completely pass us by as they fall outside the range of what we can hear, see or smell. In this chapter I shall discuss different expressions of language by animals, in the context of their social function. The aim is to present a picture of just some of the extensive current research into animal language. Many research projects have only just begun or have recently taken a different direction and are incomplete. For example, research into birdsong has been under way for a very long time and much is already known about its structure, but pinning down the exact meaning, beyond generalities such as defending territory, requires study of the context within which that song occurs, and of social relationships. This kind of research is still in its infancy.

Alarm calls

When danger threatens, we warn others. We call out 'Fire!' when there are flames and 'Hey!' or 'Look out!' when it seems a road accident is about to happen. When something is about to fall, we describe what it is and where it is coming from. Other animals warn one another too and many species do this with one or several alarm calls. For many animals, a good alarm call can literally be a lifesaver.

Scientists know a lot about how animals warn other animals about intruders, probably because alarm calls are easy to recognise. They often sound to human ears like an exclamation of fear, something like 'Help!' or 'Look out!' However, research has shown that alarm calls can have various meanings and are sometimes quite complex. Studying them teaches us not only about the communication systems of other animals, but also about how they experience and view the world.

Prairie dogs live in tunnels underground, with different rooms for sleeping, giving birth and going to the toilet. Their territory is not large and they always remain in the same area – much like humans who spend their whole lives in the village where they were born – which makes them easy prey for a number of predators. Once a predator knows where the prairie dogs are, they know that at a certain point they will naturally come up to look for food; all a patient predator has to do is wait. As a result, prairie dogs have developed a large number of advanced alarm calls, which sound to a human ear much like the twittering of birds. If you hear a lot of this chattering at once, it sounds like a bark in the distance, which is how prairie dogs got their name. Underground prairie dogs do not make much noise, relying mainly on taste to communicate. When one prairie dog meets another, they greet each other with a French kiss. This is how they recognise whether the other prairie dog is family, friend or enemy. They can be seen giving the same greeting above ground too, sometimes jumping away dramatically (if the other prairie dog was not friend or family), as if the kiss is really unpleasant.[1]

Prairie dogs use different sounds for different intruders. They indicate whether the intruder is coming from the air or on the ground as the two approaches demand different responses and it is useful to incorporate this information in a call. But that is not all: they are able to describe the intruder in detail. For a human being, they mention that it is a human, how large that human is, what colour clothes the human is wearing and if it is carrying an umbrella or a gun, for example. For a dog, they mention the size, colour and shape, but also how quickly it is approaching. Different parts of the call change meaning when the order of the elements is different, which is like a simple grammar.

They use verbs, nouns and adverbs in meaningful constructions. They can also make new combinations, such as 'oval unknown threat'. Biologist Con Slobodchikoff has been studying prairie dogs for years, decoding their language – because, according to him, it is indeed a language – step by step. In addition to the alarm calls, they also have social chattering (the meaning of this is currently being investigated) and some species, such as the black-tailed prairie dog, do the 'jump-yip', which involves standing upright and popping up into the air with their front legs up while letting out a yip. This behaviour is contagious, like a Mexican wave in a football stadium; sometimes they do the jump-yip so enthusiastically that they fall over backwards. They will jump-yip, for example, when a snake heads the other way. It appears to be a jump for joy.

The alarm calls of the American chickadee are also more advanced than you might think if you heard them, providing detailed information about birds of prey, including the length of their wings, speed and method of attack. Their name comes from the sound they make, and the most important source of information in their call is the 'dee'. For an eastern screech owl, for instance, they say 'chickadeedeedee', but for more dangerous birds they use as many as fifteen 'dees'. Closer to home, chickens have different sounds – or words – for predators that come from the air and those on land. These are not really about the animal in question, but about the manner of approach. A raccoon coming from above prompts the warning signal for an attack from the air, not a signal for raccoon. As far as we now know, chickens make more than twenty different sounds, but we do not yet really understand the meaning of most of them.[2]

Non-human primates also have a large arsenal of sounds at their disposal. The vervet monkey has different sounds for all the predators in the area. Research into their responses to different alarm calls has shown that they do not blindly react to a certain sound. When the researchers repeatedly play an alarm call (for instance, to test the different reactions to the calls for snake and bird of prey), the vervets stop reacting after a few times because the caller in the recording has proved to be unreliable. This shows that they do not react out of instinct, but make an assessment; the call passes on meaningful information and is not merely a signal that generates an automatic response.[3]

Species can sometimes understand one another's alarm calls too. The alarm call of the Campbell's monkey has syntax: elements are connected as in the structure of a sentence. The alarm call of the Diana monkey does not have this feature, but Diana monkeys may still understand the alarm calls of Campbell's monkeys.[4] And there are animals that can imitate the alarm calls of other animals. The forked-tail drongo, a small black bird with red eyes, can imitate the alarm calls of more than fifty other species. When the others fly away in terror, the drongo quickly steals their food.[5] As well as mimicking human language, parrots imitate the sounds of a large number of other animals, including alarm calls. For parrots, as for the forked-tail drongo, this ability is a source of power.

Animal alarm calls are often accompanied by, or are solely, visual signals, such as gestures, posture or facial expression, or some combination of these elements. Scent also plays an important role. Some gastropods – snails and slugs – make a sound when they are attacked, but they also use pheromones in their slime trails.[6] Research into the role of pheromones and scent in communication is still in its infancy, but we know that certain warning scents – in species ranging from bees to hippos – consist of several scents, with the proportions determining the exact meaning. African bees gather when one of them uses scent to call the others, and then they attack.[7] Such attacks can prove fatal to humans. Bees communicate using different sorts of chemical pheromones, which appear to be like words that convey information about the hive, for example.[8] The Californian thrips, a winged insect, has different alarm pheromones for different threats.[9] Thrips larvae produce drops of alarm pheromone when there is a threat of danger. The pheromone consists of two substances: decyl acetate and dodecyl acetate. As the degree of danger increases, so does the amount that is produced, and the ratio of the two changes. Larvae that pick up the signal react differently, so they clearly understand what is happening. This research shows that chemical alarms work in a more complex and detailed way than was previously believed. The thrips is most likely no exception and there are probably other arthropods that communicate in this way. Humans also use scents to communicate – it seems that romantic love is driven mainly by pheromones – but we are generally less aware of this than we are of other forms of communication.

Greetings

Humans have few alarm calls for large predators, but just like many other social animals we greet one another all the time. If a group of alien researchers were to study this phenomenon they would see a lot of variation in sounds, gestures and postures. We say 'hi' and 'hello', then sometimes stop for a chat, or just raise our hands as we pass. Dutch people may kiss once, twice or three times, on the cheek or on the lips; young people from Britain or America often give each other a hug. Others may bow or shake hands, either making or avoiding eye contact. Cultural differences in greeting can cause awkwardness – for example, when one person goes to kiss three times and the other is not expecting that many, or if one person aims for a kiss and the other goes in for a hug.

Humans say hello because they are happy to see one another and/or to reinforce their bond. Gannets, monogamous seabirds, do the same. Whenever their partner returns to the nest, they perform an extensive greeting ritual, rubbing their heads and necks over each other. The male often brings gifts for the female, such as flowers to decorate the nest or to use as a necklace.[10] Kingfishers also bring gifts as a greeting for their partners, usually something edible, such as fish.[11] The same is true of jays and crows, who bring food for their partners, choosing particular gifts. It has been discovered that these birds are able to identify with the other's perspective – they choose something they think their partner will appreciate. This means that they have a 'theory of mind' (the ability to see things from another's point of view), a quality that only humans and other primates were previously thought to possess.[12]

Humans who share their lives with other animals are well aware of the possible variations in their greeting rituals. Animals who live in the same household often greet one another, and animals usually greet familiar animals and strangers differently. Dogs can be very enthusiastic when a familiar dog or person returns home, or even when a stranger comes to visit. In greeting, dogs like to sniff to obtain information about another dog's status and characteristics. Once the other dog has been approved, play can be a good way to get to know each other better. There is not one standard form of greeting between dogs; dogs may be ignored in greeting, or looked at, some may receive a wag of the tail, and if one of the dogs is anxious or uncertain there can be growling and barking. There is often more information communicated in such exchanges than humans are able to follow.[13] For instance, dogs are good at interpreting the meaning of other dogs' growls. Research using recordings of growls has demonstrated that they know from a distance whether it is about protecting food, stopping an intruder or if the dog is angry, while to humans many of the nuances are lost. When dogs are happy, they wag their tails to the right; when they are insecure or scared, to the left. Other dogs react to this and understand that there is no problem when the tail is wagging to the right, but they become tense when the tail is going to the left. The length and position of the tail are also important.[14]

Male baboons fight and as they have sharp teeth they often get injured. They do not play together or groom one another, so greetings are actually the only friendly encounters they have. As a result, they greet one another often. This is a rather intimate affair, as it often includes allowing the other baboon to hold their penis or even put it in their mouth – a vulnerable position, particularly considering those sharp teeth. The greeting ritual works as follows: one male approaches another male, acting out threatening movements. Then they smack their lips, which indicates that they want to greet the other baboon, and they make their 'come hither' face, with their eyes narrowed and their ears flat on their heads. The other baboon usually responds by smacking its lips back and they make eye contact, which in different circumstances is a sign of looking for a fight. Then one baboon shows the other his behind, the other baboon mounts him briefly, gives him a feel and tugs his penis, and quickly moves on. Sometimes they reverse roles. This usually takes just a few seconds. Ethologist Barbara Smuts has studied the greeting rituals of baboons and states that they convey information about social status, willingness to cooperate, and age and gender. Old males often complete the greeting ritual peacefully, while with younger males sometimes one wants to greet and the other does not, and so the ritual is often broken off prematurely. Smuts thinks that greeting is primarily important for assessing others' willingness to cooperate.[15]

This not only teaches us about how baboons greet one another, but also about the function of greeting. Smuts states that we, as humans, rely heavily on language for agreements about the future, and that we therefore believe that other animals do not make such agreements.[16] But the baboon greetings show an agreement about the future. Ethologists Smuts and Marc Bekoff and philosopher of science Colin Allen[17] argue that similar kinds of social agreements are made in the playing behaviour of dogs and other species, such as wolves. (I will return to this in more detail in Chapter 6 when discussing metacommunication.) It is important to understand that a greeting is often not just a greeting: many animals are not only saying hello, but also exchanging information about their intentions. To do so, just like humans, they make use of facial expressions, gestures, body language and sounds.

Identity

A few years ago, there was big news: dolphins call one another by name.[18] Like humans, they all have a unique sound that they use to introduce themselves to new dolphins and to call one another. Dolphins are far from the only animals that have names. Parrots receive a name from their parents.[19] Squirrel monkeys have a special 'chuck' sound for each individual.[20] Bats have names that they use to call one another so that they can stay together in the darkness.[21] This can be particularly useful in a big group. A name is handy because it allows you to call someone else and to indicate that it is you who is coming.

Identity is not only communicated by voice. Hyenas live in fluid social relationships in which the females are dominant. In their interaction, they make use of scent signals from their anal glands, which occur in 252 different arrangements and form an individual profile that changes over time. The scents are also overwritten by other members of the group, which allows passing outsiders to form a good picture of both the individuals who live somewhere – their age, gender, status, health, perhaps their mood – and the strength of the group as a whole.[22] With dogs, the scent from the anal glands – every dog lover knows it well – also provides a similar sort of profile. Urine and excrement provide information about identity too. Sometimes dogs in the city, who have never met before, seem to have an inexplicable antipathy towards each other; most likely they have long been aware of the other's existence because of the traces of scent they have previously encountered, and they have some reason for hostility.[23]

Many animals make use of scent in excrement and urine. Hippos, for example, like to mark out their territory with dung, as do rabbits.[24] Lobsters have little tubes under their eyes that are filled with urine, which they spray into others' faces. The males do this when they are fighting. Lobsters fight often and remember who they have fought. They also have a mental map of who lives where. Only the strongest male mates with the females and female lobsters only mate when they have just shed their shells. They spray urine in the male's face to daze him and they dance a little. During mating, the male protects the female, but when she has a new shell the male leaves, and the next female might come along. Females do not fight one another.[25]

Like cats, snakes have a Jacobson's organ. Located in the roof of the mouth, this is an organ of chemoreception that is part of the olfactory system, which these animals use to smell. Their tongues capture scent particles that they place in the Jacobson's organ, which has two openings, allowing them to smell the world in stereo. Snakes use this to find both predators and prey, and to communicate with other snakes. The trail that their body leaves behind and the air that they pass through contain pheromones with information about their gender and age, and whether they are pregnant.[26] Young snakes follow this trail to find the location of the shared hibernation space. Puff adders, venomous snakes found mainly in southern Africa, not only leave behind scents for others to follow but also camouflage their own scent in order to deceive predators.[27] Snakes also communicate by touch, and some cobras make low growls.[28]

Wolves make use of similar scent signals to dogs. In addition, they howl. Both in the frequency and in the harmony, they give clues about their own identity and about their relationships; wolves howl or sing longer and louder to wolves with whom they have a stronger bond.[29] Their howls probably share information with each other, but we do not yet know precisely what. Coyotes sing and share information about their identity as well. The howling of coyotes is also a way to call members of their own group and to let other packs know that they are there.[30]

Dingos – Australian feral dogs that are genetically somewhere between the wolf and the dog – can both bark and howl. They bark rarely, their barking is shorter than that of domestic dogs and they sing less than wolves. Howling can be an individual matter (to discuss food or hierarchy) and because the sound travels over long distances it is a good way to communicate in the Australian wilderness. Dingos also sing in groups, as an expression of pleasure, to warn others and to communicate with other groups about the size of the pack without having to engage in a confrontation. When more dingos are singing, the frequency rises.[31]

Within species, different groups of animals sometimes have their own dialect. The songs of whales differ from group to group. Sometimes whales pick up a popular song from a certain group and it becomes a hit in that group too. Parrots live in communities of 20 to 300 animals, which all have different dialects.[32] Some parrots can speak the dialect of more than one group. The territories of the white-crowned sparrow are so sharply defined that when you stand on the border you can hear one dialect in the songs on the left and another on the right.[33]

Great tits have dialects too, and research has also been carried out into their transmission of social norms. Captive tits were taught to use a red or blue door to open a food cage containing a mealworm, a particular delicacy for these birds.

Then the birds were released into a wild population, which they quickly taught how to get the mealworm. Small trackers recorded which birds reached the mealworm and through which door. Twenty days later, three-quarters of the population understood how it worked, and the vast majority chose the door that had been taught to the first bird. When the cage was removed and put back a year later, the birds immediately started using the same door again. This is remarkable, as three-fifths of the birds from the original population had died in the meantime. The researchers believe that social norms probably also exist in other animals that live in stable social groups; behavioural innovation, passing on new skills, helps populations to survive.[34]

To determine whether animals are aware of who they are, or of the fact that they are someone, researchers have developed the mirror test. This test involves sticking a red dot to the animal's forehead and placing the animal in front of a mirror. If the animal attempts to remove the dot from his or her forehead, this is an indication of self-awareness – that is, the animals are able to recognise themselves as selves in the mirror. Elephants, magpies, chimpanzees, pigs and many other animals have been found to have this self-awareness. There are problems with the mirror test though: first, there are some animals who do not mind having a sticker on their skin. Second, in some cultures looking at yourself in the mirror is not good form. Third, it is not that suitable for animals whose other senses are more important than sight.

To start with the first point: elephants use mud to keep cool and to prevent itching, so they often do not object to a little thing like a sticker on their skin and therefore score badly in the mirror test, in spite of their intelligence and socially minded attitude.[35] We find the second, cultural aspect in gorillas, who are social animals and assumed to be self-aware, but they are naturally shy and long eye contact is not common among their kind,[36] so they too score badly on the mirror test.[37] The same applies, incidentally, to children from some non-Western cultures.[38] Out of eighty-two children from Kenya, only two passed the test, whereas Western children pass the test almost without exception – clearly the difference here is cultural, not cognitive. Thirdly, the test is also not very suitable for animals whose sight is not good. Dogs are more focused on scent than on sight, so Marc Bekoff came up with the yellow snow test, a variation on the mirror test.[39] Dogs live in a universe of scents, which inspired Bekoff to carry out an experiment in which he collected pee from the snow and investigated how his dog reacted. The dog in question, Jethro, spent considerably less time sniffing his own pee than that of other dogs, so he was clearly reacting differently to the scent profile of other dogs than to his own.

Food and love

Animals who live in groups do so because it is safer to look for food and to bring up offspring together. But life in groups also has disadvantages, such as competition in times of scarcity.

Animals who live in groups often have well-developed systems for communicating with one another about such issues. Individual ants go out looking for food at random, while ants as a colony use a food-finding system. This works as follows: scouting ants search randomly for food; when they have found some, they return to the nest, leaving a scent trail behind; other ants follow this trail to the food and leave their own scent trail on the way back, so the route to the food becomes increasingly efficient.[40] Older ants are better than younger ants at finding food and following the fastest route.

Stingless bees have an extensive repertoire of actions for communicating the location of food to one another.[41] They dance, make noise, and use complex chemical signals, which are made up of different smells, like words in a sentence. Certain species of stingless bees have a preference for chemical signals – scent traces with pheromones – from their own nests. This is learned, not innate.

Individual animals also share food with each other. Parents feed their young, but other creatures in the group may also help one another, sometimes without immediately requiring something in return. The expectation is often that others will do the same in a pattern of reciprocal altruism. Vampire bats demonstrate an intimate form of reciprocal altruism. They feed on the blood of mammals, which they search out at night. If they have not eaten for seventy hours, they die, so vampire bats who have eaten will feed their less fortunate group members by vomiting blood into their mouths. This usually occurs between relatives, but not exclusively.[42] With Rodrigues fruit bats, we find a similar attitude in females, who help one another to give birth, even if they are not related.[43]

Alarm calls are, incidentally, also a form of reciprocal altruism: the caller draws attention to itself with the aim of helping others. If all the animals in a group do the same, it makes the group safer. Callers call because others are doing it too and they are engaging in a form of collaboration.

Grooming works in a similar way, also strengthening the bonds between individuals.

Eating habits also teach us something about the memory of particular species. Chimpanzees can remember good fruit trees from three years ago and will return later to see if there is more there for them to eat.[44] Grey mouse lemurs, who eat mainly fruit, have better spatial memories than animals with a more readily available diet, such as bark or leaves. As fruit is available only in certain places and at certain times of year, the lemurs have to be better at remembering the location of their food.[45] Tits, crows and jays hide their food in autumn and remember where it is so that they can find it again in the winter.[46] Scrub jays adapt their food-catching strategy depending on who is watching: if they are within eyesight of an unfamiliar scrub jay they eat the food; if they are within hearing distance, they hide it very quietly; and at other times they will cache food in the presence of other scrub jays, depending on their relationship. This means that they use experiential projection: they understand whether the other bird will steal the food if it sees where it is hidden and they anticipate this happening.[47]

Food can also help to increase the prestige of an animal and to strengthen their position within the group. Roosters eat in silence, unless there is a female nearby. Then they call out about the food they have found in the hope of increasing their prestige. They sometimes use their food call even when there is no food around, just to lure the female.[48] This brings us to the subject of love. Showing off about food is one way to make an impression, but other birds have even more elaborate forms of ostentation.

Bowerbirds collect beautiful objects – snail shells, leaves, flowers, bits of plastic, stones that they colour with berry juice, beetles that they kill expressly for this purpose, even little blue feathers that they kill a bird to obtain – which they use to build bowers. Then they lure in the female by singing and dancing for her. The female comes to take a look; if she likes the bower they will embark upon a relationship. The females are easily distracted though, so a male might work for ages on his bower and do his very best dance, but if he is interrupted by another male, he could be back to square one.[49]

Albatrosses are monogamous and live for a long time – sometimes to the age of sixty – and so they take no risks when it comes to choosing a partner. Their complex mating ritual is an intricate form of dance, which involves an entire range of ritualised actions, such as calling, looking, pointing, pecking at feathers. Albatrosses are able to learn the rules, principles and structure of this physical language more quickly if they can copy it from older birds. Sexually mature from the time they are five, in the subsequent years' mating seasons, which last for a few months a year, they dance with many different partners, refining the dance every year, and every year the number of partners they dance with becomes smaller. Finally, after three or four years, only their true love remains. The two birds that stay together develop their own language in the courtship dance that they perform over the years, a dance that is unique to that pair. They will usually remain together for life.[50]

Caribbean reef squid have chromatophores in their skin, which means that their cells contain biological pigment and reflect light. As a result, they can change colour by tensing or relaxing the muscles attached to the cells, allowing them to take on a camouflage colour, for example, or to communicate with other squid. When a male meets a female he would like to mate with, he shows his feelings with the pattern of colours in his skin. The female then indicates with her own colours if she likes the look of him. Usually there are plenty of rivals around – other males who want to mate – and the reef squid are able to communicate with two others at the same time. The side facing the female gives her a signal – a white stripe is an invitation to mate – while the side facing the other male gives a signal that tells him to clear off. If the female responds with a zebra display and makes her body turn darker and darker, she does not want to mate. The colour patterns can change very rapidly and are rather complicated, so humans do not know exactly what information is being passed on. Some researchers believe that the changing colours have a grammar.[51] The reef squid understand one another perfectly.

Colour is also an important form of communication among fish and many species can take on camouflage colours. Coral reef fish are brightly coloured and, like the reef squid, can change their colour. They also make use of ultraviolet light (which humans are unable to perceive)[52] and communicate with the coral reef.[53] One type of parrotfish can show an image of an eye on its tail if a predator is nearby.[54] As with the reef squid, colour in fish is believed to be a complex language that humans still know little about. To seduce their lovers (and to discuss other matters) fish also make growling, chattering and 'pop' noises. They do so by vibrating their swim bladders, a gas-filled bag on the stomach.[55] All fish can hear this sound, although they cannot all make it, and their enjoyment of speaking depends on the species, and perhaps also on the individual. Gurnards are the most talkative fish; they actually make a kind of growling sound all day long.[56] Cod are not as vocally oriented and only make sounds when they are spawning, so that the male and the female do it at the same moment.[57] The bigeye, *Pempheris adspersa*, makes a sort of popping sound, which has to be interpreted by other fish like Morse code.[58]

Many animals dance to attract the attention of the object of their affections, and to show off their good points. Male fiddler crabs – small crabs with one very large pincer (around a third to a half of their bodyweight) – will attract females by standing in front of their burrows and doing a kind of dance with their big claw. They also use this large claw to engage in arm-wrestling contests.[59] Flamingos dance in a group, stretching their necks upward and taking small steps. They raise their beaks high in the air, swaying their heads to and fro. After the dance, the group breaks up into couples to mate.[60] Male lance-tailed manakins, blue birds with black-and-red heads, turn up as a group to seduce a female. They sit in a line beside her on a branch and then do a sort of circle dance, with the one who is closest to her jumping onto the next branch and then joining the line at the back, and the next bird moving up, as if on a conveyor belt. The dominant male then mates with her if she is interested; the other males are most likely there to help him in order to gain his favour, or because they hope that they will one day become the alpha male.[61]

There are, of course, animals who use their voices to impress a possible lover. Male pandas make a noise that resembles the bleating of a sheep when they like a female, and the females react with a sort of chirping bird sound if they are also interested. Sometimes the result is panda babies (who say 'wow-wow' when they are a little unhappy and 'gee-gee' when they are hungry). Recent research in zoos has demonstrated that the chance of pandas having babies is much greater when they are able to pick a partner themselves, rather than when people make that choice for them on the basis of genetic selection.[62] Chinese researchers are working on a panda translation machine in order to understand these animals better. It is hoped that more understanding will also help them to protect pandas.[63]

For many spiders, the ritual of love is a pretty delicate affair. Before the males embark on their quest, they have to fill up their pedipalps – appendages on the front of their bodies – with sperm. They do so by first depositing a drop on their web and then sucking it up. Then they have to find a female. The males have chemical sensors on their front legs with which they detect pheromones that females secrete when making a web. Once they find a female, they need to chase off any rivals and so male spiders sometimes break the webs so that other spiders cannot reach the female. If other males are already there, they will start a fight. If they are successful, they then need to let the females know that they are not a prey that has happened to get caught in the web but a male of their species who wishes to mate, so garden spiders strum a rhythm on the web,[64] while wolf spiders and jumping spiders, which can see better, do a dance. Wolf spiders often bring a gift, too, in the form of a wrapped-up prey (or a shell if they are really keen to mate but have been unable to find actual prey).[65] Females may then respond that they also wish to mate, or that they are not in the mood, by running away or by causing the web to move. Sometimes the males try anyway, running the risk of being eaten by the female.

Conflict and mixed messages

Nature films make it seem as though the animal kingdom is a violent place, but aggressive signals are often simply intended to chase off the opponent, thus resolving the conflict. Fights are risky because you might get wounded and medical attention is generally not available, so most animals would prefer to avoid actual conflict. This means that most aggressive communication is about bluffing or chasing away, rather than about challenging the other to a fight.

In conflicts, we make much use of language, and the same applies to other animals. Lizards communicate with their bodies: for example, there are species that do a kind of push-up in front of each other until one of the two darts away. Bats fly at one another and use complex combinations of sounds, which the other bat can respond to with an even more complex arrangement. Guinea pigs chatter their teeth. Rhesus monkeys have five sorts of aggressive calls. Others, such as macaques, have different sorts of aggressive calls.[66] Closer to home, cats will occupy a particular place or stare in order to chase away another. Many people will not even notice that one cat chased away the other, but for the cats this is a very meaningful interaction.

In *The Expression of the Emotions in Man and Animals*,[67] Charles Darwin developed the principle of antithesis, by which he meant that the expression of certain emotions indicates different ends of a spectrum. A dog that is angry will make itself look larger, assuming a threatening posture and issuing low growls or barks. A dog that is scared will make itself small and adopt a submissive attitude.

Although these postures do occur, many messages from animals are more ambiguous than Darwin suggests. Many uncertain dogs go on the attack, whereas dominant dogs can assume an apparently relaxed posture until the other dog pushes it too far. A dog can make itself both larger and smaller at the same time, as in the 'play bow', which is meant as an invitation to other dogs. It is not only the individual postures that convey information; changes in posture can show that tension is building or ebbing away.

Some biologists argue that Darwin's principle can also be applied to the sounds that animals make: low sounds, like growling, represent anger and dominance, while high-pitched sounds, like whining and squeaking, are fear. Slobodchikoff writes that there are certainly cases when it works this way, but this is not always what happens. Many companion animals understand that an angry, low-pitched human voice means dominance or punishment and sweet, high-pitched sounds are encouragement. Humans also perceive low-pitched human voices as more dominant; in every American presidential election since 1960, the candidate with the lowest voice has won (with the exception of Al Gore, who actually won the popular vote, but not the Electoral College). There are people who believe that discrimination against women is primarily a question not of gender, but of voice, height and other physical characteristics that inspire confidence in human beings. That those factors inspire confidence in us is, of course, connected to ideas about gender.

Wapiti deer and red deer are similar, but make very different sounds. The wapiti calls at very high frequencies, ending with a sound like nails scratching a blackboard.[68] Red deer roar, making a low, rolling sound, and adapt the frequency to the size of their opponent – the larger the opponent, the lower the sound.[69] The white-nosed coati has a lower sound for aggression and a series of higher sounds for friendly interactions, but there is a wide range of harmonic variation between those two poles, which cannot be explained simply by assuming that low-pitched sounds always mean aggression and high-pitched sounds are always about friendliness or fear.[70] That picture is too simple, and makes the messages of animals appear too straightforward.

The difference between communication and language

Many people make a distinction between communication and language, and believe that animals are capable only of the first, while humans can do both.

According to Slobodchikoff,[71] animal behaviourists see communication as a closed system with three components: a sender, a recipient and a signal. In this closed system, everything happens on the basis of instinct; animals respond in a pre-programmed way. When a prey animal sees a predator approaching, for example, they freeze; when the predator comes closer, they run away. The distance of the predator generates an inbuilt response. Language, on the other hand, is an open system that offers different options for the question and the response, both in terms of an animal's inner world and the outside world. An animal can deal creatively with the situation presented and make a meaningful choice.

In all vertebrates – including long-extinct species such as Neanderthals – the FOXP2 gene, more commonly known as the language gene, has been found.[72] It is not only responsible for language, but is also linked to forms of learning. That's not to say that invertebrates do not have a similar gene; in animals that have evolved differently we sometimes observe that the body, through evolution, has found a different answer to the same question. One example of this is the brains of birds and humans. In terms of evolution, birds split away from mammals millions of years ago, and yet both species are capable of similar reactions. The brains of birds are not much like mammal brains, but they are able to generate the same kind of intelligent reactions.[73]

From an evolutionary perspective, it would be strange for humans to have language and for other animals not to have something similar, for there to be a hard border between language and instinct-based communication. Darwin already argued that the difference between humans and other animals is of degree, not kind. Marc Bekoff, who researches emotions, morality and justice among animals, follows this view, and argues that 'if we have something, they have it too', which is to say, if humans feel love, joy, sadness, then so too do other animals – not in exactly the same way, but along the same lines.[74]

Slobodchikoff argues that the basis of language is a meaningful signal being passed from one animal to another within a certain context. This can be learned or instinctive, or both, and we see it in humans and other animals. In humans, facial expressions, such as a smile, are, for example, innate, but they can also be learned within a culture in certain situations. As we saw earlier, signals from animals are not random but are often ordered according to certain rules of syntax – as in, for example, chickadees, chickens, bees, lizards, wolves and prairie dogs. Various animal languages also have a grammar. From an evolutionary perspective, this makes sense: all kinds of animals use language to integrate, categorise and organise information, and it is important to do this as efficiently as possible.

In the 1960s, the linguist Charles Hockett came up with thirteen criteria that language must satisfy in order to be language. These criteria are still referred to in discussions of the language of animals.[75] The first six characteristics can be applied to most communication systems, not specifically to language, and there is no debate about the fact that they are found in the languages of other animals too. These are: a sensory system for receiving and sending information; the ability to broadcast and receive signals; the ability to make signals that disappear so that new signals can be sent; the capacity to understand the signals of others of your own species; the ability to hear your own signals; and a system that is specifically for transferring information. There is debate however about whether or not the next characteristics are applicable to non-human animals. Characteristics seven and eight are about meaning: they concern semantics, the meaning of a word; and arbitrariness, that a word is not a reflection of whatever it refers to, but an abstract symbol. Then there is the characteristic that symbols come in discrete units (such as words) and that the symbols themselves come in units (such as syllables). The remaining characteristics dictate that in a communication system it has to be possible to make new words; the transmission must be cultural or traditional; and the communication system must be capable of conveying information about events that take place elsewhere or at a different point in time.

Cultural transmission has been demonstrated in a number of species. Many birds learn songs from their parents. The dialects of groups of animals also give some insight into this. A language consisting of units that are in turn composed of other units applies to prairie dogs and to certain bird languages. That sounds have meaning is evident from the discussion of alarm calls; it is not clear to what extent this meaning can be arbitrary and refer to abstract symbols, but the possibility certainly exists. In communication with primates, and with Alex the parrot, it has become clear that animals are capable of making new words or word combinations for new objects and situations – and the prairie dogs do exactly that with their 'oval unknown threat'. While this is not proof that new word combinations occur regularly in animals' species-specific languages – this is something we do not know, yet – it does show that they have the capacity. The ability of an animal communication system to focus on the future or the past, or to discuss a place that is elsewhere, is demonstrated by the research into greetings and playing behaviour, and there are indications that whales and elephants exchange such information.

To further explore whether animal languages are language, Slobodchikoff points to contemporary discussion about recursion as an important characteristic of language. Recursion in language takes place when new sentences are placed within sentences to add additional meaning. For example, in the statement 'Eva says that elephants use low-frequency sounds to communicate information' the sentence 'elephants use low-frequency sounds to communicate information' occurs within a larger sentence. Some linguists consider this the most important characteristic of human language. As the example of the elephants shows, it is quite possible that this also applies to the language of elephants and it has certainly been demonstrated in the language of various birds.[76]

A second characteristic that Slobodchikoff adds is efficiency. This is about how precise a language is. If you have a word for a concept or an object that indicates exactly that thing, you are more precise than if you devote an entire paragraph to the description of something and can only indicate its vague outlines. He says that other animals are very good at this. One single alarm call from a prairie dog, for instance, which lasts a fraction of a second, can tell listeners that they need to take cover because a hawk is diving in to grab them. All that humans are capable of doing is shouting something like 'Look out!' or 'Above you!'

The examples discussed above are, of course, not watertight proof that there are animals with a language that meets all the criteria – far from it, in fact. But the research, though still at an early stage, does show that animals communicate, that they do so in a more complex way than we previously believed and that certain characteristics in different species correspond to human language. This casts doubt upon the exceptionality of human language, and raises the question of what exactly language is, and who gets to decide. Maybe there are characteristics that certain animal languages have that human languages lack. I am not sure we will ever truly be able to understand the nuances of communicating via colour patterns or by chemical scent signals. Definitions of language decided by humans will always favour humans, and so we should include other characteristics when engaging in thinking about other animal languages. The study of the above characteristics is not pointless, however, as it can give us insight into the structures of their languages and possible parallels to human language. This in turn can give us insight into their social interactions and their lives, and can help us to develop other research questions.

CHAPTER

3

LIVING WITH ANIMALS

Within three years of intensive training, Chaser the border collie learned the names of 1,022 toys. Her vocabulary is larger than that of a three-year-old human. Not only can she bring them on request, she can also divide them into categories: the balls with the balls, and the dolls with the dolls. She understands that words refer to objects, that certain verbal instructions refer to objects, and that the names can refer to objects and to categories.[1] Chaser has a good memory; John Pilley, her trainer and companion, had to write the names on the objects in order to remember them. She can also make deductions: when she hears a new word she will be able to locate the object it refers to by ruling out objects she already knows the names of. This research resembles studies of linguistic proficiency that attempt to make animals speak in human words but the difference is that there was no attempt to make Chaser repeat the words – it was not focused on teaching her human language. The words that Chaser learned are tied to objects, so the experiment is defined by different criteria: comprehension means comprehension within the context of fetching and categorisation of objects, not the learning of abstract concepts.

After spending three years learning words – which Pilley got fed up with; he thinks that Chaser could easily have learned far more words – Pilley and Chaser started working on grammar. Chaser understands sentences with simple grammar.[2] When she is asked to take the toy giraffe to the leopard, she does so. When asked to take the leopard to the giraffe, she does that too. Previous research had already demonstrated that dogs understand simply constructed sentences – something that most lay people are aware of too when their companion animal responds to 'Fetch the ball!' or 'Come here' – but Chaser intuitively demonstrated understanding of this kind of grammar, and Pilley developed this further by providing rewards. Pilley believes that Chaser's talent for grammar is in part connected to her breed: border collies have been bred to work with sheep and so have to be focused on humans while still keeping an eye on the sheep. However, although border collies may pick up this sort of knowledge more quickly, Pilley assumes dogs of other breeds are also capable of doing the same. Chaser is not the only dog with a large vocabulary. Rico, a German border collie, learned three hundred words and could categorise objects.[3]

The relationship between dogs and humans is a special one because they evolved together. That long shared history and the process of domestication have bonded dogs to humans, and vice versa; they became part of each other's culture. Dogs started barking in order to communicate with humans, and humans have learned to listen to that barking and take meaning from it.[4] When humans listen to recordings of barking dogs, they can tell what mood the dogs are in – even if they themselves do not live with a dog companion. They can also understand what the dog wants from the intensity of a recorded growl. When shown photographs of parts of human faces, dogs are able to assess how the person in the picture feels, and voices are even easier for dogs to interpret. This is largely a result of domestication; dogs can read human gestures and expressions far better than wolves, their wild relatives. If a human hides food under a cup and puts two empty cups down beside it, a dog in an experiment will follow the human's instructions; if the human points at the cup on the left, the dog will sniff that one first. Wolves tend to ignore the human and use their noses instead. All kinds of variations on this experiment have been done. Wolves also lack this sensitivity to pointing and physical instructions from humans even if humans have raised them, which indicates that it is related to domestication and not a matter of training.

Humans who try to explain canine behaviour on the basis of wolves' behaviour – something that is popular among amateur dog trainers – often forget that domestication has changed not only the body, but also the psyche of dogs. This mutual focus is not only something that we – dogs and humans – have learned by living together: thousands of years of shared history have also had a genetic influence. The best example of this is a recent study that showed that when a dog and a human who love one another look at each other, they both produce oxytocin – the cuddle hormone that humans produce when they see or hug a loved one.[5]

The biologist and philosopher of science Donna Haraway points out that dogs actively participate in processes of domestication, helping to shape its outcome both from a socio-cultural point of view and on an individual level. Haraway uses her own relationship with her dog, Cayenne Pepper, as an example. Cayenne Pepper is part of her life and the things they do together strengthen their bond and their shared world. They do agility training together, a sport that requires not only the dog to learn all kinds of new skills, but also the human. This has changed Haraway's perception of the world around her, not because she has started to resemble a dog herself, she jokes, but because new knowledge and experiences enrich our world view. When we learn something new together with a dog, the dog influences what happens, which in turn influences our own view of the world. Haraway emphasises the physical and material character of this interaction: working and moving together with the dog changes humans' bodies and minds. Humans are not just 'brains on sticks'; we react to pheromones and oxytocin and send signals with our physical reactions.

Words also play a role in forming shared worlds with other animals. Dog trainer and philosopher Vicki Hearne argues that when we teach a non-human animal a word, the world of that animal and of the human becomes larger.[6] She refers here to Wittgenstein and writes that when we learn a new language game, 'we learn to read the darkness'. This does not mean that a dog and a human understand a concept or a word in exactly the same way. While humans mainly use their eyes for orientation, dogs use their noses. Dogs see less well than humans – about six times less; they are not colour blind, but they do see fewer colours – but they can smell about a thousand to a million times better than humans. These are estimates, as the exact figures are unknown and the ability to smell varies between breeds – dogs with longer noses are better at smelling than pugs or bulldogs, for example. While we orient ourselves by sight, dogs make a map of smells. They can make out the individual scents within compounds too. When we smell pea soup, the dog smells carrot, leek, pea and the other ingredients. In order to bring about understanding, a human has to bear this in mind. When a dog and a human go tracking together, each of them experiences their environment in a different way. The human navigates on sight; the dog uses his or her nose. However, they are both working on the same project and actions gain meaning through experience and practice.

Learning a language game enriches the world of animal and human, and Hearne writes that it enables animals to communicate with humans in a more complex way, for example when a human teaches a dog to fetch. Hearne taught Salty, a pointer, to fetch a dumbbell. Teaching her this language game allowed her to express herself more fully. She can fetch other objects, and take the dumbbell to someone else, but it also gave Salty an opportunity to show her creativity and the opportunity to make jokes – once when she was asked to bring the dumbbell Salty brought the lid of the rubbish bin instead.

Hearne sees a hierarchy in the learning of language games: the human decides what is learned and how. However, the precise form of a language game in which animals of different species participate is not predetermined. Rather it is a dialogue in which a human or a dog begins, the other participant responds, the first reacts to that response and so on. Dogs are not passive recipients of information but can influence the shape of the interaction through their actions. This process has no fixed end; shared understanding can continue to grow between humans and dogs who have been together for years.

Humans and other animals are born into a specific social context. This environment shapes us, and we shape the environment. One of the ways in which we do this is by using language. This is how we learn to understand ourselves and the world around us. We also use language to exert influence on others. The German philosopher Heidegger saw language and the world as 'equiprimordial', equally original.[7] By this he means that there is no language before the world and no world before language. A world only develops because we express ourselves and give it meaning; the fact that there is a world means that we can express ourselves and give meaning. Heidegger believed that non-human animals are incapable of language because they cannot understand themselves as a self in the world. He based this on the research of biologists at the time, particularly Jakob von Uexküll, who thought that animals were all fixed within their *Umwelt*.[8] This environment is determined by the situation animals find themselves in and by their senses, and is therefore different for every animal. A spider perceives the world as a spider and can only think a spider's thoughts. According to Heidegger, humans are the only animals that can transcend this. They are able to think about the world beyond their immediate experience of it, and do this in language. But the stories above show that the reality is far more nuanced: other animals make sense of their surroundings in their own languages. Whether humans truly comprehend themselves as humans is also questionable.

Heidegger even wrote that, as they have no concept in language for death, animals cannot die; they simply disappear. This might seem plausible; animals do not leave a will and do not indicate to us in language that they know they have an awareness of their mortality or the abstract concept of death. But if we use this logic it raises the question of whether we humans are able to die. We know that someone who is dead will never return, that the body stops living, but this knowledge does not solve the larger mystery of life and death. We do not know exactly what death is, which is why stories about what happens after death are so appealing. Animals express themselves in different ways from humans, but there are similarities in the way that they attach meaning to relationships, learn through communication to understand themselves and the world, and at the same time play a part in shaping that world. Crows, elephants and other animals have mourning rituals and are interested in others of their kind who have passed away, and many animals will keep watch over the body of a dead group member. We perhaps do not understand them well enough yet to assess the value and depth of this behaviour, but it would be premature to deny that they have an understanding of death.[9]

Domestication

Domestication is described as a multigenerational relationship in which one group of organisms significantly influences another group's reproduction for their own benefit. Various theories exist about how and when exactly domestication of different species of animals by humans began and much of the evidence is open to multiple interpretations. It is generally believed that dogs' wild predecessors – wolves, or a common ancestor – were attracted to human settlements between 11,000 and 32,000 years ago because of the availability of food – human excrement is in this context an important food source. Humans saw the advantages of their presence – for example, as guard dogs – and encouraged them to come closer. The canine individuals who were friendlier towards humans stayed closer, mated with one another and had even friendlier offspring. Each generation became increasingly attuned to humans. There are different opinions about who initiated this process. Some believe that humans domesticated dogs. Others think that dogs came of their own accord, domesticating themselves through natural selection. There are even some who argue that it was the dog who domesticated the human and that this relationship is also responsible for human language development, as humans needed to call their dogs, but this interpretation is contested.[10]

If we compare domesticated animals with their wild relatives, we see that the domesticated variant has retained more of its infant characteristics, such as playfulness, friendliness towards strangers, an urge to discover, big eyes, flappy ears, a large head in relation to the body and a greater ability to adapt to new situations. This phenomenon is known as neoteny. We see it when we compare dogs and wolves, or bonobos and chimpanzees, or humans and their ancestors. The theory of evolution sometimes assumes a survival of the fittest. However, Darwin pointed out that co-operation, empathy and collaboration are necessary for the survival of many species. It pays to be friendly. The ability to adapt to changing circumstances is also an important characteristic from an evolutionary point of view. Some scientists therefore believe that certain species, such as dogs, domesticated themselves before humans began any breeding programmes. The same could apply to human beings, who had to adapt in order to function in increasingly larger societies, losing some 'wild' characteristics in the process.[11]

Animals who have been domesticated – or who domesticated themselves – are often dependent on the presence of humans for food and care to a greater or lesser extent. This requires far more human–animal interaction than with wild animals. Throughout their shared history, humans and domesticated animals have become attuned with one another. Irrespective of our individual knowledge of animals, we as a species, and in our culture, have become involved with other animals. Humans often think that other animals are part of nature and humans are about culture, but animals have their own cultures and are sometimes part of human communities, and humans, just like animals, are bodies and are part of nature. Haraway uses the word 'naturecultures' to describe the interconnectedness of nature and culture.[12] Domesticated animals show us that animals are part of our culture too, and there are different varieties of domestication. The meaning of concepts such as nature and culture also change over time. Some refer to the present era as the Anthropocene, the age determined by humans. This may make it seem as if almost everything is about culture, but at the same time nature often demonstrates that we as humans are nature too – illness, for example, may force us to confront our physical existence, and earthquakes may remind us that we are always part of a bigger world.

Here I am! Where are you?

British naturalist Len Howard thought that the way in which birds were studied in her time, the 1950s, distorted reality. Birds were studied in laboratories, in experiments that were designed to be repeated. The philosophy behind this type of research is called behaviourism. Behaviourism draws on methods from the natural sciences and focuses on predicting and controlling behaviour. In this approach, human and non-human minds are studied as black boxes, of which the content will forever remain a mystery, and is irrelevant: only outward reactions that can be measured have scientific value. Descriptions of acts should be avoided. Wild birds are scared of humans, Howard noted, so life in a laboratory causes a lot of tension; nervous birds react differently from relaxed birds and this has an impact on the results of the research. They also cannot fly, or have social contact, which influences the outcomes of the studies.

Howard, who was not a biologist but a viola player and a bird lover, decided to take a different approach. She bought a piece of land near Ditchling, to the south of London, with its house, Bird Cottage. Her plan was to make the surroundings as safe as possible for the birds who lived there so that she could study them on the basis of trust. She literally opened up her house to the birds, who could fly in and out through its windows. Birds are naturally curious and the birds who lived nearby soon came to take a look. Howard made places where the birds could build nests and she fed them. They realised that they were free to fly around the house, and it was not long before they occasionally started to perch inside.

Howard wrote two books about her life in Bird Cottage,[13] describing the characters and lives of individual birds. She was particularly interested in great tits, many of which became very tame, but she also discussed a large number of others. She initially wanted to study birdsong, but soon found that their individual personalities and relationships were worth studying too. She emphasised the individual intelligence of birds and spoke out against the idea that birds act only out of instinct, as was commonly believed at the time. Howard described both the communication between the birds, which she saw as very complex, and their communication with her in detail. When she spoke to them, the birds reacted to the smallest variations and inflections in her voice, and appeared to understand her well. One example was their communication about butter. The birds loved butter and often came to beg for it. They would sit beside her plate and look closely at her face. If she had a friendly expression, they took a step closer, and if she spoke encouragingly, they would help themselves to a bit of butter. If she sternly said 'no', the birds took a step back. An even sterner 'no' would result in them stepping further back, while an angry 'no' would make them fly back out of the window. If she called them back, they came closer again, but more tentatively than before because she had spoken to them angrily. New birds soon learned what she meant.

Konrad Lorenz also lived with a large number of birds and other animals.[14] He thought this was the only way to study animals properly. He raised most of the birds himself, so that they were tame and saw him as family. Howard's work showed that it is not necessary to have birds from a young age in order to tame them; Lorenz believed that it was. His most extensive, lifelong study was with geese. Geese communicate with one another in many different ways: with their voices, by making different sounds; with gestures; body postures; rituals; scent and so on. There are also different relationships and encounters between geese and humans. Very close bonds can be created with geese raised by humans – for example, if the human calls the goose, he or she will stop flying and come straight down. This affection prompted Lorenz to call geese the best companions after dogs. Friendships are also possible with geese that were not raised by humans, but they will remain more distant. Geese treat humans differently in different situations. They also recognise interactions with other animals, such as dogs, and objects, such as cars, with the context once again playing a role. They can learn that a dog on a lead is not to be feared, but to take care when one is running free, or that a familiar dog is not a danger while unknown dogs might well be. Geese also understand when humans imitate their call – for example, to warn of danger – and humans understand the body language of geese.[15]

Like Howard, Lorenz practised what is now called narrative ethology and in his work he wrote biographies of different geese and their relationships with one another. In narrative ethology, individual stories say something about the bigger picture. Lorenz also carried out more standard biological and ethological research. However, the design of Lorenz's research was very different from Howard's. As a human researcher, Lorenz allowed himself to be influenced by the geese but was still the one who decided the rules of the game. He caught young geese, decided which group they would live in, and when they had to be where. Howard, on the other hand, effectively gave up her life with humans in order to dedicate herself to the study of birds, leaving the birds to make their own rules. Though there was a great intimacy in both cases, Howard demonstrated that captivity and domestication are not necessary to achieve this and that it is possible to study non-human animals on the basis of trust and freedom.

Alone or together

Cats and humans perceive the world differently. Cats can smell and hear better, and their sight – which is worse than humans' during daylight but better than humans' in the dark – is focused mainly on movement, because cats were originally hunters. Scent plays an important part in a cat's perception of their social environment. Urine and excrement provide information about who has been in the area and serve to mark out territory; cats deposit their scent by rubbing their heads on objects and other animals. Cats have a scent organ at the back of the mouth, and if they think something smells interesting, they will exhibit the flehmen response, curling back their lips and drawing in the scent. They often greet each other with a raised tail – and sometimes greet their human in the same way. Cats have developed specific ways to communicate with humans, the most important of which is meowing. Kittens make sounds to call their mother, but adult cats do not meow to one another, only to humans: a skill that cats have taught themselves through interaction with humans. Cats are, then, bilingual.[16]

It is often thought that cats do their own thing. Dogs listen to humans and can be taught all kinds of commands, but cats would rather sleep all day.

Dogs live in a pack, but cats are solitary creatures. There are of course all sorts of differences between dogs and cats, which affect their relationship with humans, but the image of cats as showing little interest in their social context or humans is incorrect. Studies of cats in homes and in shelters where they live in groups demonstrate that cats do see themselves as part of the pack. Cat researchers Janet and Steven Alger argue that we can best understand how feline communication works by studying interactions between humans and cats.[17] They believe the best method for this is ethnographic study, which maps out a community within its own habitat – in other words, not in a laboratory.

The Algers studied the social structure and culture of a community of cats at a shelter, focusing on the symbolic interaction between cats and between cats and humans, the ability of cats to see things from another's point of view, and how feline norms and values are established.[18] The cats had all been neutered and there was enough for them to eat. As a result, fights connected to food and mating rituals occurred only to a very minor extent. The humans working in the shelter viewed themselves as part of the cat community and only rarely assumed the role of the human who knows best. When feeding, the cats' preferences and social relationships were taken into account. Friendships between cats were respected in the allocation of space – who slept where – and in adoption, so that friends were housed together. There were cats in the shelter who liked to live with another cat, and there were cats who slept in baskets in groups of seven or eight, washing one another and eating together. There were cats who liked just about everyone but who were only tolerated by other cats. In general, the cats preferred to sleep together, even if there were sufficient solitary places available. The situation in this shelter was certainly not natural, but then most cats form part of human communities and cat communities, by sharing homes, urban environments and, as in this case, animal shelters.

For a long time, the emphasis of research into animal behaviour was on aggression and the defence of territory, but the Algers' study of feline friendships and communication shows that affection and cooperation often occur between cats. There are several explanations for this emphasis on aggression; for example, that such behaviour might be easier to recognise and evaluate than animals sitting calmly next to one another. Feminist philosophers of science also point out that an emphasis on dominance and rank occurs primarily among male researchers, and for a long time the majority of researchers were men. This has influenced the image of certain species. Chimpanzees, for example, were long thought to be very aggressive, but characteristics such as empathy were not studied.

The Algers also studied symbolic interaction between cats and humans in common households: how cats and humans create symbolic worlds through a process of interpretation in which interactions between individuals create meaning and shared understanding. This takes place on different levels. Cats can solve problems – they can work out how to open doors and windows, but they also know how to get human help when necessary. The Algers cited the example of a cat who had her collar stuck in her mouth and asked her humans to free her by coming up to them for help.[19] Cats make informed choices about, for example, whether or not to go outside in bad weather and food – sometimes they wait to eat in case something tastier comes along. Memories play a part in this; previous situations influence their choices. There are differences between individual cats in temperament and ability to learn. Cats and humans have shared routines, influenced by both parties; a cat can demand play, for instance. Symbolic interaction is not limited to that between cats and humans but also occurs between cats and dogs.

The cohabitation of humans and cats influences individual relationships and lives. Cats and humans develop shared habits – going to the toilet together, walking the dog together, going to sleep together. Living as part of a human household also influences how cats live with one another: humans may adopt cats and then simply deposit them in the same area as other cats. Research shows that neighbouring cats will coordinate their territories and the times when they go out in order to accommodate one another. Cats stay in their own area and if they share a territory they go out at different times of day. Cats who share a home usually tolerate one another and may go out and explore together. City cats often like to plan their activities according to their human's daily rhythms. Farm cats who depend on catching their own prey are active at night because this is the best time for hunting – when the mice and rats are active – whereas house cats often sleep at night along with their primates. Living with humans also has an effect on prey animals in the surrounding area. Cats who live off prey usually kill instantly, while house cats who are not hungry may play with their prey for a long time. Some cats do not hunt at all. Researchers believe that cats may get out of the habit of hunting completely when they have been domesticated for a long enough time.[20]

Sharing space

Julie Ann Smith lives with rabbits.[21] She wants to give the rabbits that she looks after on behalf of an animal welfare organisation as much freedom as possible while also taking good care of them. She views this as a conflict: as a human, she determines the space they can use yet she also wants to respect them as individuals with their own wishes. She tries to provide as much freedom as possible for the rabbits within the available space and she aims to find new forms of cohabitation. One of the ways in which this happens is very literal: through the use of space. The rabbits run free through the house. Smith has covered up plug sockets and the like, so they are in no danger. She describes how the rabbits always used to make a terrible mess in a certain room in the daytime. She would tidy it up in the evening, but the next day the rabbits just started again. This continued until she discovered that the rabbits were following a system: they love tunnels and hiding places and they were organising the room accordingly. When she realised what was going on, she saw the appeal. This was a way for her to communicate with the rabbits, and she uses this example to demonstrate the importance of experimentation when living with other animals. Even when the boundary is clearly defined – in this case, the animals' captivity in a house as they are probably better off there than outside – there is space within those limitations for the rabbits to act and to make choices.

In many areas, cats are free to leave the house and to return when they want to. Outdoor cats sometimes find a new family or eat in different households. Dogs, though, particularly in Western countries, are kept in the house and on a lead. In her own household, writer Elizabeth Marshall Thomas investigated how to increase the freedom of dogs in an urban environment. Misha the husky, who was staying with her, would jump over the fence every night and take himself for a walk. Sometimes he would bring along his girlfriend, her daughter's dog, Maria. Marshall Thomas followed Misha to find out what he was up to and realised that he knew where he was going and had found his own ways to deal with certain situations. For example, when crossing a busy road, he used his ears and not his eyes. Maria did not know the way as well as

Misha, and when she got lost she looked for humans, expecting them to take her home, which is exactly what happened.[22]

Marshall Thomas lived with a group of around eight dogs, and her human family. The dogs raised one another; she did not have to toilet train the puppies, for example, as they learned from the adult dogs. At a certain point she moved to a house with a large garden, which she fenced off so that the dogs could go outside whenever they wanted without being able to run away. The dogs gradually grew closer to one another, spending less and less time in the house. Marshall Thomas sometimes went to sit outside with them because she felt the need for contact. One day she discovered that they had dug a hole in the ground as wolves do (one of the dogs was a dingo) and that they spent quite a lot of their time there. She concluded that given the choice dogs need one another more than humans.

Ted Kerasote also looked for ways to increase the freedom of his dog companion Merle.[23] He lived in a small village in Wyoming where most of the dogs roamed free. Before he allowed Merle to do the same, he first had to teach him a few things: don't hunt livestock because the farmer will shoot you, don't hunt big game because it can be dangerous and be careful around traffic. When Merle had got the hang of these constraints, Kerasote made a dog flap in his back door, which Merle was free to come and go through as he chose. He often spent the daytime with his friends in the village, both dogs and humans. He liked to go for a walk with Kerasote at fixed times, and he always came home on time to eat. He also slept at home every night. Sometimes he would bring a girlfriend home. Although there were risks associated with this way of life, which are absent from the life of a dog whose movements are always controlled, Kerasote believed that Merle's life was so much richer. Kerasote writes that as he became more independent, he also became more intelligent, better able to cope with challenges, to think for himself and to communicate with others, including individuals of different species.

In order to function, domesticated animals must learn the language of their own kind, as well as the language of humans and of other animals in the household or village. Marshall Thomas shows that dogs can find their own way in a society that is determined largely by humans and that, in certain circumstances, they may prefer a life with other dogs. Kerasote demonstrates that humans can encourage dogs to become more independent and that dogs and humans can find new ways to live together. The dogs themselves play a crucial role and exhibit different temperaments. Through new forms of cohabitation, new forms of communication can develop. The reverse also applies. Shared language games can provide a means of thinking about new ways of living together and forming communities, or support ideas about these subjects. What these examples show is that species does not determine the quality of the communication or the closeness of a relationship. Some humans form relationships with other animals more easily than with members of their own species, while others prefer the company of other humans – and the same applies to animals. There will be humans who connect with one another in many ways, but there are also humans who have little in common, while there are animals and humans who do share a great deal. It is quite conceivable for a human and her dog to have more in common – preferences, understanding, how well they know each other, how they respond to certain events – than that person does with a random neighbour. All sorts of connections are possible – the likelihood of a shared language is not dictated by fur or a tail.

Cooperation and resistance

Companion animals are not the only ones who were domesticated. Human relationships with animals such as cows, sheep, horses, chickens and pigs also date back tens of thousands of years. These relationships have changed with the development of agriculture as an industry. These animals were once part of the social life of farming families and of the rural and urban landscape, but they have increasingly disappeared from sight.

There are some historical examples of large-scale animal husbandry. The Egyptians bred animals to sacrifice in the form of mummies, and captured wild animals for the same purpose.[24] The Romans selected breeds of chicken that laid a lot of eggs and kept them in large numbers.[25] But this is nothing like today's industrial farming – the techniques that are used these days to make animals produce as much as possible are new, and the numbers of animals that are bred and killed for food and other animal products are incomparably greater.

This increase in scale and the industrialisation of animal husbandry clearly has consequences for the animals' relationship with humans, and for interspecies communication. On most farms, agricultural animals are no longer part of the farmer's family – this would be practically impossible with a million chickens – and the relationship instead aims only to ensure the animals function well and are as profitable as possible.

Communication among farmed animals has also been disrupted because they live so close together and lack stimulation – the stories about chickens pecking one another to death and pigs eating one another's tails are well known. Both species are very social animals with an extensive language.

Pigs recognise one another primarily by their scent and have a complex series of vocalisations, which humans have not yet mapped out fully. Their social bonds are like those of elephants, and their prefrontal cortex – the part of the brain that has been implicated in planning complex cognitive behaviour, personality expression, decision making and moderating social behaviour – is enlarged, as it is in humans and other primates who hunt and forage for food. They root in the ground, investigate the world around them with their snouts, and wag their tails when they are happy. They take good care of their children, have empathy for others, are playful and have a good memory.[26]

In addition to their large arsenal of alarm calls, chickens make use of sight, touch and scent to communicate about present, past and future.[27]

They are good at counting – chicks are better at doing sums than human babies[28] – feel empathy and jealousy, and individual chickens have been shown to have very different personalities.[29] Mothers communicate with their chicks when they are still inside the egg, later adapting their life lessons according to their offspring's ability to learn.[30]

Sheep are known for their docile natures, but in reality they are creative animals with good memories, and they live in complex social networks. They use all kinds of sounds, body language and pheromones to communicate.[31]

Much of the communication between grazing animals is subtle. Cows[32] and horses[33] communicate, for example, with a lot of eye contact and ear movements; researchers are only now working to map these two forms of communication. Not all interaction is this subtle though. Once every so often, a cow or a pig escapes on their way to the slaughterhouse. There are also occasional stories in the news about farmers and their families being trampled or attacked by farm animals. The historian Jason Hribal has carried out extensive research into acts of resistance among domesticated[34] and wild[35] animals. Hribal points out that we assume that humans built the economy although other animals have actually played a very important role. He even suggests that we should see them as members of the working class.[36] They are, however, unreliable workers; it takes a lot of energy to keep them in line and Hribal argues that the resistance of working animals has in part led to them being replaced by machines, and thus helped instigate the Industrial Revolution. Animals escaping from abattoirs or resisting human control also influence public opinion, and sometimes legislation. Hribal uses the example of the United States Camel Corps, a mid-nineteenth-century experiment by the US army in using camels as soldiers. When the camels resisted in all kinds of ways, including screaming, spitting and biting the human soldiers, the humans they were supposed to be working with began to loathe and fear them. They could no longer work together and the practice was stopped. The army claimed that it had just been an experiment, but it was the camels who turned it into an experiment by refusing to cooperate.

In modern interactions between farm animals and farmers, there is still a strong focus on dealing with – and curbing – resistance. If you visit a pig shed, you will be told not to stand with your back towards the animals. Sheds, milking machines and transporters have been designed in such a way that animals have as little room as possible for resistance. The Australian philosopher Dinesh Wadiwel says that resistance is a good lens through which to look at animals as it reflects their creativity and their willpower.[37] This can be seen even if we just study human responses to that resistance. He writes, for instance, that the resistance of fish can be seen in the mechanisms that humans have devised to catch them, such as rods and hooks.

Resistance is a type of communication that can take on very different forms, in humans and in other animals. In *Fear of the Animal Planet*, Hribal describes the resistance of wild animals in circuses, dolphinariums and zoos. By presenting a large number of accounts of animals escaping, injuring or killing their keepers, sabotaging activities and destroying objects he demonstrates that acts of resistance are not isolated or rare. They are often presented as such because the humans who run and profit from these institutions do not want the animals to appear unhappy. The most famous example of animal resistance is probably Tilikum the orca, who was portrayed in the 2013 documentary film *Blackfish*. Tilikum, kept at SeaWorld Orlando, killed three people: two trainers and a man who trespassed in his enclosure. There is convincing evidence that he did this on purpose. There are other orcas in captivity who have injured or killed people, although there are no known instances of orcas murdering humans in the wild. Nearly all orcas in captivity have physical and mental complaints. Ninety per cent of the males, for example, have a collapsed dorsal fin caused by stress, which does not occur in the wild. Tilikum himself was probably depressed and maybe psychotic.[38]

Resistance can also take place on a small scale. Anthrozoologist Leslie Irvine says that play between humans and other animals can be a form of resistance,[39] because in small everyday activities power structures are revealed. As the idea of taking the subjectivity of other animals seriously is considered strange in our society, actually doing so in practice, in a game, for example, is an act of resistance. For a game to succeed, one has to take the individuality of the opponent seriously. Irvine points out that creativity is important in play and that individuals reveal their personalities by playing. Animals have their own preferences when playing, and play goes through a development. Play gives humans the opportunity to get to know their animal companion better, and vice versa, and playing with animals gives adults an opportunity to do something just for fun. By taking animals seriously and not seeing the species boundary as an obstacle to meaningful interaction, humans and animals who play together can serve as an example for more sceptical humans, who are invited to view animals differently.

Back in 1580, the French philosopher Montaigne wrote that when he played with his cat he did not know if he was playing with her or if she was playing with him.⁴⁰ What is clear is that both players play, and that this is necessary for the game to take place.

CHAPTER

THINKING WITH THE BODY

Hans was born in Germany at the end of the nineteenth century. By the time he was four years old, he could divide and multiply, and calculate roots. Not only did he have a knack for maths, he could also spell words, read, give the time and date, differentiate between musical tones and intervals, and recognise colours.

Hans, not a human child but a horse, answered the questions that people asked him by tapping his front right hoof on the ground. Wilhelm von Osten, his human, started appearing before audiences with him in 1891. Before long the wonder horse had attracted the attention of the press, with more and more people coming to see his performances. Some were convinced that Hans was a genius while others were sceptical. To investigate whether there was any deception involved, the government set up a committee led by the philosopher and psychologist Carl Stumpf and consisting of thirteen experts in the field of equine intelligence, including a veterinarian, the director of a zoo, a horseman and a number of teachers. In September 1904, their conclusion came: no tricks or fraud were involved. Because they still did not fully understand Hans' intelligence Stumpf asked an assistant, Oskar Pfungst, to continue the investigation.

Pfungst researched Hans' capacities thoroughly, first without an audience and without von Osten. Hans answered the questions just as well as usual, ruling out deliberate deception by von Osten. Then Pfungst investigated whether Hans would also give the right answer if he could not see his examiner, and if the human did not know the answer. In these cases, Hans was unable to answer. Pfungst suspected that Hans reacted to minuscule changes in the questioner's body language. When he took the place of the horse himself, he discovered that his questioners all unconsciously made a small head movement when it was time for the last tap of the hoof. The conclusion was that Hans was no wonder horse. This study led to the practice of double-blind behavioural research, with the researcher not knowing who belongs to the test group and who is part of the control group; or more generally not knowing the desired result, so that he or she cannot accidentally influence the test subject. The risk of this happening applies equally to studies of humans as to studies of other animals, as Pfungst himself demonstrated. Von Osten, incidentally, continued to perform with Hans, and the audiences kept coming.

The Belgian psychologist and philosopher of science Vinciane Despret, who has carried out much research into the relationship between animal researchers and animals, points out that Hans was indeed an intelligent animal, but in a different way than originally thought.[1] Hans could read the smallest changes in human body language. Horses are good at communicating with humans, but normally do so primarily by touch (as when a human rides a horse) and not so much by sight – but Hans was able to understand visual signs. Hans also trained the humans who questioned him; the longer they worked with him, the clearer their signals became, although they were unaware of this themselves. Through body language, some conscious, some unconscious, horse and human learned how to read each other and become increasingly attuned.

The example of Hans raises all kinds of questions about thinking, cleverness, animal research and the role of our experience, factors which are connected. Animals are often considered to be completely different from people, and therefore more difficult to know. However, there are some problems with this view, both regarding the capacities of non-human animals, and how we get to know others, human or non-human.

Psychology and animals

Ancient Greek philosophers such as Aristotle and Plato tried to fathom what knowledge is and how we acquire it. Until the twentieth century – when experimental psychology was developed – thinking about thinking occurred mainly in philosophy. In the first half of the twentieth century, behaviourism was very influential in thinking about thinking and in animal research. The most important proponent of behaviourism was the American psychologist B. F. Skinner.[2] Behaviourism views psychology as the scientific study of behaviour. Even inner events, such as thoughts and feelings, can be approached as behaviour. The aim is to predict and control behaviour, rather than to describe and explain it, and the focus is on connections between behaviour and environment, not on a deeper origin or under-lying social structures, unless these are immediately apparent.

The linguist and philosopher Noam Chomsky is one of the major critics of behaviourism.[3] He argues that behaviourism is unable to explain certain phenomena. For example, children who learn a language are able to understand and reproduce far more sentences than can be explained by a model that assumes all linguistic proficiency is directly learned in detail. One of Chomsky's arguments is that humans have an innate capacity for language, which explains structural similarity between different languages around the world. This is the hypothesis of universal grammar: we are born with language already within us and it only needs to be exposed. According to Chomsky this ability is purely human; no other species of animal possesses it. The experiment with Nim Chimpsky discussed in the first chapter was developed to demonstrate this distinction. Chomsky's theory, also known as generative linguistics or generative grammar, is rational, not empirical: we cannot empirically demonstrate this linguistic ability but only discover its existence through the study of language, with our intellect.

Chomsky's work inspired the branch of philosophy known as cognitive psychology. In behaviourism, the inner world, or the brain, of humans is conceived of as a black box: only the result of processes that occur within that box can be measured and is relevant to science. Cognitive psychology, however, focuses on the contents of that black box, our inner world. During the development of these concepts, the computer served as material for comparison;

scientists looked for information systems and information processing in the brain. Study of the brain plays an important part in obtaining information about information. The physical aspects of cognition are studied in the neurosciences through experiments that make frequent use of test animals – everyone has seen pictures of monkeys with electrodes on their heads, or their skulls opened up – as experimenting on humans is not considered ethical. So the brains of animals serve here to gain insight into human brains. Nowadays cognition is more generally studied in cognitive science, which is based on interdisciplinary insights from psychology, philosophy, linguistics, neuroscience and information science, with a focus on researching the mental processes and intelligence of humans and other animals.

Behaviourist research into animals is still carried out, and the picture that Chomsky paints of language as belonging only to humans still has many followers. In animal research, non-human animals are still often viewed as very different from humans, and many experiments are designed accordingly. Animals are kept in laboratories, where measurable results count, and the researchers do not form a bond with them as this might influence the results. In recent years, however, animals are increasingly being viewed as subjects, which has consequences for how they can and should be studied.

Baboons

Barbara Smuts studied baboons in Kenya and Tanzania for twenty-five years.[4] The group she got to know best was the Eburru Cliffs Troop, a pack of around 135 baboons that moved around in an area of seventy square kilometres. Smuts travelled with them every day for two years, from sunrise to sunset, though she did sleep elsewhere. She saw no other humans for the first few months, but later slept in a camp with other researchers, with whom she had only limited contact. At the beginning of her study, Smuts tried to get closer to the baboons in order to gain a better understanding of their behaviour. She walked towards them, stopped when they moved away from her, waited until they relaxed again, and then headed towards them again. This strategy resulted in little progress. After some time, she discovered the signals the baboons gave one another when she came too close – mothers called their children, others gestured to one another – and she learned to stop before the tension became too high and the baboons ran away. Once she had identified this behaviour, she was soon able to get up close.

The baboons grew used to her presence. Researchers call this habituation, which means that the baboons, or other animals who are not accustomed to humans, adapt to the human who has come to observe them. For Smuts, though, the opposite happened: she had to adapt in order to fit in with the troop and the baboons simply went on with their own lives. During Smuts's doctoral research, her supervisor taught her that researchers need to be as invisible as possible, but Smuts learned something different from the baboons. Baboons are social animals and, in their language, ignoring each other is seen as a provocation; only friends can ignore one another when they meet. So when a baboon approached Smuts, she soon learned that it was better to make brief eye contact or to give a little growl than to ignore the baboon. If she followed baboon etiquette, the baboon went on with what she or he was doing; if Smuts ignored the baboon, she was sending a signal that the baboon did not understand, and this created tension. If she showed that she had seen the baboon and did not mean any harm, the baboons accepted this as a sign of respect. Through this mutual understanding and by participating in the communication,

Smuts learned a great deal about greetings, personal space and communication that she would never have gleaned otherwise.

The baboons learned that Smuts was a calm person who meant them no harm, and Smuts learned to move like a baboon. She felt vulnerable, but was confident that she would understand when a baboon was angry with her, as she had learned to read their behaviour. The baboons may have accepted her for the same reasons. By moving with them, she slowly learned to perceive her surroundings like a baboon. One example was her reaction to changes in the weather. In the rainy season, storms on the savannah can be seen approaching from far away. The baboons became restless when a storm was coming, but they still wanted to go on eating. They knew exactly when it was necessary to find shelter from the storm, so they could keep eating for as long as possible. For months Smuts wanted to seek shelter long before the baboons stood up. However, at a certain point, just like the baboons, she knew exactly when the right moment came, without being able to explain why. She simply knew.

Living in a group taught Smuts more about the baboons. When the group finds a place with enough mushrooms for everyone – mushrooms are a rare delicacy that the baboons will fight over – they give cries of joy before they all start tucking in. She also twice witnessed a ritual in which the baboons sat around small pools of water, quietly gazing into them, before moving on to their sleeping place. Smuts knows no other examples of such behaviour in the scientific literature and saw this as a kind of mystic ritual, wondering if she was perhaps witness to something that non-human animals normally do not allow humans to see. Smuts also learned something about truly being part of a group. As humans, we are not accustomed to coordinating our movements with those of others, nor are we used to attuning our routines to nature, or the earth, as the baboons did during the storm. In the group of baboons, Smuts experienced what it was like really to be part of something that was bigger than herself. She also began to see her body and her thoughts differently: as a primate belonging in a group of primates.

As in the study of baboons' greeting rituals described in Chapter 2, Smuts carried out serious scientific research, in which she made use of video recordings to capture the smallest changes in behaviour. Her personal experiences with the baboons, however, also significantly informed that research, giving her insight into aspects of baboon society that other researchers do not have. Anthropologist Matei Candea studied researchers working with meerkats and says that close interaction with the animals you study provides a kind of insight that cannot be found in the canon of science.[5] Like Smuts, he demonstrated that the animals train the researchers just as much as the other way around. Rather than avoiding contact with animals, which is often impossible anyway, including activities such as the interaction with the researcher within the research data is a way to improve our knowledge. Despret, who pointed out the same process in the case of Clever Hans, agrees. When researchers learn to read animals and a sort of common language develops – as in the case of Smuts and the baboons – they gain a richer perspective on these animals and more insight into their lives. This has consequences for the research: if we consider animals as subjects with their own view of the world, we will ask them different questions and we will ask them in a different way.[6]

In the 1960s, Jane Goodall carried out research into chimpanzees in Gombe,[7] and gave those chimpanzees names. She also referred to them not as 'it' but as 'he' or 'she'. A great number of scientists considered this unacceptable humanisation of the chimpanzees. Goodall's research was of great importance as she was the first to show that chimpanzees use tools, when it had been assumed that the use of tools was what distinguishes humans from other animals. This discovery showed Goodall's critics that her methods were perhaps a little unorthodox but that she was, in fact, making an important contribution.

Although some scientists are still wary of anthropomorphism, we have seen a definite move towards viewing animals as subjects in recent years. Not attributing any thoughts and feelings to other animals is not a neutral stance, and has since been termed 'anthropodenial'. Many of the characteristics that humans think are solely human are of course found in other animals too. An example concerns the writings of Lorenz on love. Lorenz often used human concepts to describe the behaviours and emotions of the animals he lived with, which attracted a lot of criticism. When he wrote about romantic love between animals, he was accused of anthropomorphism, but this has since been demonstrated elsewhere.[8]

In animal language research, communication is often a prerequisite for the success of the research. We find this in Irene Pepperberg's work with Alex the parrot. In order to be able to research Alex, Pepperberg had to discover how they could achieve mutual comprehension. With Alex, words were helpful; for Smuts and the baboons, communication was mainly about eye contact, gestures and body language. This might seem less scientific than double-blind studies in laboratories, but such studies are also based on certain assumptions – and, in some cases, prejudices – about animals.

Phenomenology

Barbara Smuts's research not only tells us something about the best way to study animals, but also places the role of experience at the heart of getting to know someone else. Thinking is often conceived of as something that takes place inside the mind, but this implies a separation between body and mind, and a separation between thinking and the world. This view is challenged in phenomenology, a twentieth-century philosophical movement in which the experience of phenomena is central. In contrast to empiricism (which assumes all knowledge comes from experience) and rationalism (the idea that reason is the only source of knowledge), phenomenology focuses on the essence of perception. According to phenomenologists, an experience is always focused on something: we do not simply see at random; we always see something. This focus on something in the world is called intentionality. In phenomenology, because of the emphasis on experience, thinking is always necessarily connected to the world, to perception and to experience.

The French phenomenological philosopher Maurice Merleau-Ponty argued that thought is always embodied.[9] He pointed out that the body is not simply an object like other objects in the world. It is not comparable to a table. We do not *have* a body, we *are* a body. If we touch our left hand with our right hand, the right hand is the object of the touch, while we can simultaneously feel that the hand itself can also feel. The fact that we can feel ourselves as a feeling subject is what makes us a physical self. The body makes experiences possible, and perception – the way in which we acquire knowledge – is in the first instance a physical activity, not a cognitive one. In our body we also carry the past with us. Previous experiences that have been recorded in our body ensure that we perceive the world in a particular way. Habits too are primarily physical; by developing a habit we add actions to the repertoire of our body, which enriches our daily existence.

For Merleau-Ponty, language is also embodied. It is often thought that we form thoughts in our head and then speak them. Perhaps it works that way in an interview or while writing a book, but normally, when speaking, we are not expressing previously formed thoughts. Language is a physical activity: we think by speaking. The words we speak complete our thoughts and make them our own. Words are part of our body's toolkit; Merleau-Ponty also calls them 'ways to sing the world'. With our bodies we understand others, and language and speech connect subjects to one another and to the world.

Another phenomenologist, the German philosopher Martin Heidegger, investigated what *Sein*, or Being, actually is,[10] which he saw as the key question of philosophy. Heidegger described a number of characteristics of how we are placed in the world that are also important for thinking about and with animals. Firstly, we are situated. This means that from our birth – from when we are, as Heidegger says, 'thrown into the world' – we exist in a context, being shaped by it and helping to shape it. We cannot occupy any perspective outside ourselves; our ideas and thoughts do not exist in a vacuum but are coloured by our experiences. Secondly, on an existential level, we are always with others. Heidegger does not mean that we are not alone – we are, in fact, alone at the same time – but that the structure of our Being is a Being-with-others. This becomes clearer when we look at what he writes about language. Heidegger saw a strong connection between language and what he called 'world' – not the planet Earth, but our 'lifeworld'. We express ourselves in language, thereby forming this lifeworld, and at the same time we understand it through language. At one point, he refers to language as the house of Being. Through language we can understand ourselves as selves; without it we would be stuck in immediate experience.

Both Merleau-Ponty and Heidegger saw humans as different from other animals. For Heidegger, there was a fundamental difference, because people can understand themselves as Being and, according to him, other animals cannot. For Merleau-Ponty, humans and animals are connected because we all exist as bodies, but there is a difference between the types of experiences humans and other animals have. Although Heidegger and Merleau-Ponty did not attribute certain characteristics to animals, such as language for example, and Heidegger exaggerates the rationality of the human, their theories are still illuminating in thinking about animals, precisely because they emphasise physicality and being in the world.

Wittgenstein's lion

The later work of Wittgenstein can also be classified as phenomenology.[11] In his early work he looked for fixed, immutable principles in language, until he realised that language cannot be defined in this way. I have discussed the importance of

Wittgenstein's ideas for thinking about language with animals, and here I want to look at an element of this, the importance of social practices. When philosophers speak about Wittgenstein and animals, or even about language and animals, they often cite the following quote: 'If a lion could speak, we could not understand him.' What they mean is that animals are so different from us that even if we had a common language we still would not grasp what they meant. But that interpretation of the quote is incorrect and also indicates a poor understanding of Wittgenstein's philosophy. First, Wittgenstein is not making a statement about animals here; the lion is merely an illustration. This becomes clear if we read what precedes this section in the text: Wittgenstein writes that humans can be a mystery to each other, which we may notice when we visit a foreign country. Even if we bring a dictionary we may not understand the people as we do not recognise ourselves in their body language, practices and habits. Words alone are insufficient to bridge the gap. Wittgenstein then introduces the lion as very different from a human. It is significant that he did not mention a dog, cat or any other domesticated animal.

Vicki Hearne also discusses this quote, and argues that Wittgenstein is exaggerating the difference of the lion here.[12] A lion trainer would understand the lion very well – in fact, a lion and their trainer already speak a common language. Although I agree with Hearne that it is an exaggeration to describe a lion as completely different from us, it is also important to look at Wittgenstein's underlying point, namely that we struggle to understand others when they belong to an entirely different culture, one with which we are unfamiliar. Wittgenstein is saying that language is connected to our way of life and receives meaning only in a context, through particular activities. If we want to say something meaningful about the language of others, then we need to study the practices within which that language is used. If others, animal or human, are incomprehensible to us, it is not because their mind or thoughts are inaccessible. It is because we are not familiar with their habits and manners and the other things that lend meaning to living together. This also works the other way around: when humans and other animals share lives, households and habitats, understanding deepens.

Doubt and knowledge of the other

Some philosophers think we can never really know anything for sure. This position is called philosophical scepticism and there are a number of variations on the theory in the Western philosophical tradition. We find the first sceptical thinkers among the ancient Greeks. Pyrrho of Elis (360–275 BC), who is seen as the first representative of scepticism, believed that as all of our assumptions are based on other assumptions, we cannot know anything for sure and we constantly need to reflect upon our judgements. There are good arguments for different positions, and so it is better to suspend our judgement than to take a position.[13] Descartes, who aimed to arrive at certain knowledge through radical doubt, placed modern scepticism on the agenda. With his study of the foundations of thinking, he opened up the question of whether knowledge is possible at all. Descartes saw a separation between mind and body, between intellect and passions. In thinking, we can think that we exist; beyond that, nothing is certain. As previously discussed, according to Descartes animals cannot think because they do not speak, but his ideas about thinking and knowing others have also been influential in the case of humans.[14]

Solipsism is a position that is related to scepticism and which was enabled by Descartes' separation of body and mind. A solipsist assumes that there is only one consciousness – namely one's own – and that this is the only fact of which we can be certain. It is possible that there are people around us with a consciousness like our own, but they could just as easily be advanced robots; or we could be being deceived by a god who is playing a game. The entirety of history could have been made up. This may not seem very logical, but perhaps it is illogical only because we are so accustomed to the world around us as an outside world. Solipsism is hard to prove or to dismiss, as it is not possible to prove our existence to another person beyond all doubt. Just think about it.

Language plays an important role in sceptical considerations about being able to know others in everyday life. By using language, we can give others fairly accurate information about ourselves and discuss a wide variety of subjects. Humans often assign human language a higher status than animal languages. While human language can play an important part in understanding others, it can also deceive, and there is no reason why belonging to a different species should hinder understanding or knowledge of the other. In fact, only being sceptical in the case of other animals and not in the case of humans is problematic.

Sceptical arguments are hard to refute. Wittgenstein does so by referring to the public character of language – language does not receive meaning inside one's mind, but between people. But my main aim here is not to refute scepticism.

My point is simply that theoretically being sceptical about other animals and not humans is a form of discrimination that is based on stereotypical views of animal minds, and animal languages. Non-human animals usually don't express themselves in human language, but there are various forms of communication, or shared interspecies language games, that enable understanding between humans and other animals. We often know what they mean, and vice versa. Because language is not just located in the mind, but rather it is embodied and grounded in social practices, the idea of minds as enclosed spaces to which others have no access is untenable.

For anyone who believes that humans are the only animals who experience doubt, consider that macaques would rather miss a turn in a game than make the wrong choice.[15] Researchers taught macaques to judge the number of an array of dots on a computer screen. They could choose either 'd' for 'dense' or 's' for 'sparse'. If they got the right answer, they were given something to eat; if the answer was wrong, the game was paused. However, they could also choose a question mark, which meant they did not receive any food but they also did not have to wait for a new task. In cases of doubt, the macaques always chose the question mark.

What it is like to be a bat

In a famous article, the American philosopher Thomas Nagel wondered what it is like to be a bat.[16] He wrote about this subject not because he actually wanted to understand what being a bat is like, but simply used the example of a bat as an illustration in an argument about consciousness. According to Nagel, it is not possible to reduce mental states entirely to the physical, as those who believe we are our brain may think. This does not acknowledge the fact that our experiences have a subjective character, and so it does not explain our consciousness. We are not exemplars of our species; you experience something as you, and so the same pain, for example, might feel different to you than to another. Bats make use of echolocation to communicate and to orientate themselves in the way that humans use vision. We can imagine what it is like to use echolocation and to be able to fly, but this does not mean we know what it is like for a bat who grew up as a bat to experience the world. Even if we were gradually to change into a bat, we would still lack that bat-specific knowledge. This thought can be extended to other experiences: we can imagine that a bat feels pain, but we still do not know what it is like to feel a bat's pain.

Although Nagel is right when he says that we

can never know for certain what it is like to feel another's pain – whether human or bat – we can certainly empathise with others. We can also learn to read others by watching their behaviour, interacting with them, or reading about forms of communication, and thereby gain a better understanding of how others feel and why. We can imagine what it is like to be a bat and wonder if or how it is different from imagining being a different human being; for instance, being a man if you are a woman, or simply being another individual. Imagining what it is like to be someone else is not just a question of thinking: empathy and creativity are also important, and you can get to know someone and gain insight into their consciousness through experience. This does not have to be a question of all or nothing. Maybe it is impossible for us to imagine exactly what it is like to be able to smell as well as a dog, and what kind of consequences this has for the dog's experience, but this does not mean we are unable to picture it or that we cannot understand the dog at all.

Phenomenological horses and dogs

So how clever was Clever Hans? If we measure according to human standards, or use mathematical and musical ability as the only expression of intelligence, Hans was not very bright. If we want to discover universal human grammar in him, we would not get very far on the basis of his linguistic proficiency. If we consider his brain as a black box, maybe he is a little clever: after all, he taught himself how to perform certain tasks upon receiving minimal signals from humans, even though he may not have understood what he was doing.

Derrida wrote that the philosophical tradition denies animals the possibility of response,[17] first because it was thought that they could not answer and only reacted, so were automatically excluded from consideration, and second because the questions posed were tailored to the human. Hans's intelligence was measured according to human standards and so he was considered to be not very clever. We do not know how smart he was by equine standards as the research focused on his interaction with humans. What we do know, however, is that Hans was a quick and promising student of human–horse communication. He rapidly learned to read humans and could also teach them to give him better signals with the body language they had at their disposal. Horses communicate with their bodies in many ways – they can turn their ears by almost 180 degrees, for example, and use the position of these ears to tell one another where food can be found or if there are predators nearby – and the body played an important part in Hans's thinking and communication.

Smuts, the baboon researcher, has also written about the role of the body in creating understanding between humans and other animals, and about how living with another animal can create a common knowledge. When she adopted her dog companion Safi, Smuts assumed that she, like the baboons, was an individual with her own view of life.[18] She tried not to train Safi, but instead communicated with her as an equal, using body language, words, gestures and facial expressions. She always spoke to her, particularly about things they were both interested in or disagreed about, such as eating and walking. When Safi did something she did not want her to do, Smuts would tell her, and the tone of her voice and the words she spoke were enough for Safi to understand what was expected of her. In some situations, Smuts determined what would happen – in a busy city, for example – while in others Safi would lead the way – when they went hiking or camping in the mountains. Paying attention to each other in this way meant that Smuts and Safi formed a close relationship and developed daily habits and rituals, such as morning yoga. Merleau-Ponty wrote about habits that they take place primarily at the level of the body. They make our lives richer; a new habit in one's life adds a layer to our existence.

Their interactions changed both human and dog. Their world expanded, with language playing an important role in the process. Smuts described the interaction she had with Safi from the level of her experience, and of Safi's experience. This is a different starting point than researching animals' responses in experiments or attempting to chance upon a truth about animals by thinking about them. The emphasis on reciprocity is important here: rather than acting as a human following a preconceived plan for relating to an animal, Smuts always looked at what Safi did and adjusted her actions and judgements accordingly.

Wittgenstein wrote that many problems in thinking are based on misunderstandings in language. To avoid this, we need to look at how language is used. Different approaches to animals can be seen as language games that provide a range of knowledge. In communications with animals both within and outside of research, animals were long considered and studied as objects. This was the dominant language game, and it has long obscured the possibility of other ways of thinking about animals, not least because it provided results that only confirmed the image of animals as objects. Smuts's research shows that alternatives are possible, and that these alternatives can offer us new insight into old questions. Her move to intersubjectivity gives us new options for researching experience, not only the experience of non-human animals, but also that of the researcher, and of the two together.

CHAPTER

STRUCTURE, GRAMMAR AND DECODING

The brain of the octopus is only small. Most of their nerve cells are in their arms, which can taste and touch and operate independently of the brain. You could say that octopuses think with their arms, and that those arms form a stronger connection between the 'self' and its surroundings than they do in humans.

Cephalopods – a class of marine molluscs that includes the octopus, cuttlefish and squid – are believed, on the basis of brain and behavioural research into memory and the ability to learn, to have a consciousness.[1] Certainly, various species of cephalopod are highly capable. Octopuses are known for escaping from their tanks in laboratories at night, often to eat fish from a nearby tank, sometimes returning to their own tanks of their own accord. They also use tools such as coconut halves and jam jars to hide behind or inside.[2]

The skin of certain cephalopods is a magical organ. This skin contains chromatophores that allow the creature to change colour by tensing and relaxing its muscles, which makes them masters of camouflage but also allows them to create dramatic and rhythmic displays of colour. These complex colour patterns are how cephalopods communicate extensively with others at great depths. Their appearance is an element of their behaviour. Not only the changing colour and texture of the skin, but also their posture and movement are part of how they communicate.

Research into the similarities between the signals of Caribbean reef squid and human language by the biologists Moynihan and Rodaniche has shown that the colour patterns of the reef squid are comparable to the languages of birds and primates in terms of structural complexity.[3] This communication satisfies a number of criteria that we consider specific to human language. The colours appear, for example, to be able to refer to aspects of the outside world. Different signals can refer to the strength, range, precision and the specific content of the messages they want to convey.[4]

The structure and complexity of animal languages is a relatively new area of research. Animals were long assumed to communicate with one another only through independent utterances, so little research has been carried out into sentence structure.

An exception is birdsong, where there has been fairly extensive research, although this has yielded little information about the meaning of bird languages.

Structure

Grammar is the set of rules governing the structure of language. Ferdinand de Saussure, the Swiss linguist who laid the foundations of modern linguistics, made a distinction between *langue*, the underlying structure of a language, and *parole*, the actual linguistic utterances of individual speakers. When we study language, he suggests, we need to look at the *langue*, as language usage is too prone to change. Words come and go, but grammar remains unchanged. Together they make up what language is. They cannot completely be separated either: language usage confirms the structure of language, and utterances acquire meaning against the backdrop of the structure of a language.

De Saussure distinguishes between two aspects of words: the signifier, the manifestation of the sign (such as the sound or the letters on the paper), and the signified. The signified is the mental concept to which the signifier refers, not to be confused with the physical object in the outside world (this is known as the referent). According to Saussure, words gain meaning within a language, not from the outside world. The word 'cat', for example, has nothing to do with a cat and does not acquire meaning from actual cats in the world, but through its difference from other words, such as 'rat' and 'fat'. According to Saussure, when we study language, we must therefore focus on how signs relate to one another, not on what they refer to in the outside world.[5]

Structuralism, a movement in the social sciences that built on Saussure's ideas, assumes that underlying structures of society influence humans in many different ways, rather than the other way around. Structuralism became popular in different fields of study, such as linguistics, anthropology, psychology, and even history in the 1960s and 1970s. In these fields, the focus moved from studying human acts to the fixed underlying structures that shaped them. Although structuralism is no longer very popular – no hard underlying structures that determine everything have yet been discovered – certain aspects can still be found in different fields of study, such as animal language research. This has certain risks. If we focus solely on underlying fixed structures that govern language or behaviour, we gain a sort of mechanical understanding of our object of study that leaves little room for freedom or creativity – in animals it led to a focus on instinct rather than intelligence.

It was long thought that other animals acted only out of instinct, as if pre-programmed. According to this concept, all communication between animals has a specific framework, rooted in the animal itself, with all responses fixed. In this model, the language used by animals is simple and does not allow for creativity; it mainly involves isolated reactions to events that occur. The dominance of this image of animal language is partly a result of the academic disciplines within which it developed: animals are studied in biology and ethology, which are primarily concerned with defining characteristics of a certain species, based on certain predetermined standards, not with the implications for what we consider language to be. Slobodchikoff, who argues that animals do have language, has worked to connect linguistics and animals, pointing out the importance of empirical research into the structure of animal languages. He considers Chomsky's universal grammar as a kind of internal language structure of human and non-human social individuals. Such a structure can be found in social species of animals, he writes, as all social animals encounter similar problems in dealing with our environment. He also sees evidence for this in the discovery of a language gene that can be found in the DNA of all vertebrate species.[6]

The picture that Slobodchikoff paints of language is problematic – language is more than an innate system and if we study it only empirically we miss certain dimensions of meaning, in humans and in other animals. Empirical research can teach us a great deal about the complexity of the language of other animals, but if we are to interpret it we need to rethink what grammar and language are. This is also a philosophical question.

The grammar of birds

Birdsong is the most extensively researched animal sound. Bird sounds are generally divided into songs and calls, such as alarm calls, with songs being more structurally complex and having different functions. Birds sing with their voices, but can also use their feathers, wings, tails, feet and beaks to make or distort sound; communication can include the tapping of the woodpecker or the rustle of wings. The voice organ of birds is the syrinx, which is located at the end of the trachea – the windpipe – and can make sound without vocal cords. Muscles make the cartilage and membranes vibrate, which produces sound. In a large number of songbirds, the syrinx can produce different sounds at the same time. As well as singing loudly, birds can also sing very quietly, in a kind of whisper.

It was long assumed that birds sing only to court a female or to defend their territory. There was no further consideration of the content of the songs, which were believed to have a closed structure – the birds only sang according to fixed patterns. Research into recursion – the occurrence of a construction as part of itself, or the addition of new sentences as elements within sentences – in starlings shows that the situation is not that simple. Starlings are capable of understanding new recursive additions to their language, which means that their language, like human language, is open.[7] The sentences that starlings sing are therefore not pre-programmed: there is room for meaningful additions that make new sentences. Other research, though, doubts whether starlings' grammar is really that mutable and works like human grammar. Although the answer is disputed, the question is open and meaningful.[8]

Slobodchikoff describes the complex structure of the aggressive songs of the blue-throated hummingbird. Quite a lot is known about their syntax. Five different kinds of note have been identified and classified as C, Z, S, T and E. The C is a very short phrase in which four notes sound simultaneously. The Z and S are longer trills with different frequencies. The T is a burst of sound, and the E is another short phrase with four simultaneous notes, with a different range of frequencies from the C. These sounds are combined in different ways: sometimes they begin with the Z, followed by S, T, E and again S, T; other songs begin with C, then S and T together, followed by T and then E. The songs can contain eighteen different notes that are used in different combinations. As with the starlings, this is an open system in which there is room for new possibilities with new meanings. We know little about that meaning though; such research would also require study of the context in which the songs are used. We do know that these are songs to do with territory, and it is likely that the messages reflect the birds' intentions, ranging from 'clear off' or 'come on if you dare' to 'I know where you live'.[9]

The American chickadee is, as previously mentioned, named after the sound it makes. This sound, 'chick-a-dee', is used in social contact with other birds, in discussions about territory, in quarrels and to challenge other birds. But representing the sound as a simple 'chick-a-dee' does not do it justice; the song contains a grammatical structure that can convey a great deal of information. The 'chick-a-dee' call can be divided into four components: a brief whistle, an even shorter whistle that goes up and down, a very loud sound and a longer one that sounds like a bark. All combinations are possible, and some sounds made with the songs – such as wing clapping – have meaning too. Recursion has also been identified here: elements can be repeated in a very long series. Chickadees also make a sound known as a 'gargle', which they use in conflict situations. This is a very complex sound, which lasts no longer than half a second but consists of a series of whistle-like notes. They also make certain gestures to accompany that sound, as humans do when we want to emphasise our words. The gargle can be made up of thirteen different notes, which are arranged in patterns (just as letters and syllables make up words). Eighty-four gargles have been distinguished so far. If a conflict continues, the chickadee adapts the sound by making the series yet more complex. The gargles have a structure and the birds adapt them to circumstances.[10]

In the call of the Carolina chickadee, which is also made up of four elements that vary according to context, order has been shown to be important: researchers changed the sequence in experiments and then the birds did not react.[11] Humans can also tell the difference between nonsense and meaning if the sequence of words is incorrect. Among blue-throated hummingbirds, the males and females sing differently. The song of the males is divided into five categories, which are used in different combinations, while the song of the females is much more variable and more complex, which means that we still understand little about it.[12] The song of hummingbirds has not yet been studied extensively; however, the more that research is done, the more intricate the structures prove to be.

Grammar and context

Grammar is normally seen as the body of rules and principles for speaking and writing a language. Wittgenstein also thought that meaningful language usage was bound by rules, but he used the word 'grammar' for the broader network of rules that determines whether words are used meaningfully. So, for him, grammar consists not of technical instructions for learning and correctly using a language, but rather grammar expresses meaningful language usage. Here again there is a strong connection between language and practice. The meaning of language cannot be seen independently of the ways in which it is used, and grammar must take this into account.

This way of thinking about grammar is also relevant when studying animal languages. If we study birdsong only by looking at the structure of the songs and the calls, we can get an idea of how the structure works, but that is not sufficient for us to learn how to understand its meaning or that we also need to take the context into account. Like humans, birds learn which meaning goes with which sounds. Certain situations require certain social rules. Researchers have thus far concentrated mainly on the structure of the song, in some cases in combination with brain research, which teaches us that birdsong is more complex than we believed but says little about its meaning. In order to learn how to understand the subtleties of birds' interaction it is not sufficient to classify their sounds, which could actually make the situation seem more mechanical than it is. Song analysis should always be accompanied by the study of social connections, as well as other actions and practices. The research of Len Howard or Konrad Lorenz in which people lived together with other animals offers an interesting background and angle in this context.

Chemical and visual grammar

Honeybees dance in order to explain things to one another. They also make use of chemical signals. Their dances can have two shapes: a circle and a figure of eight. When they dance in a circle – which is known as the round dance – it is to indicate that food is nearby; as the other bees can smell their way to the food, they have no need to give further instructions. When the food is more distant, the dance changes. This is because bees make honeycombs in the nest in which the queen's eggs and food are stored; these eggs hang vertically in the nest, which means the bees cannot simply point to where the food is located. Instead, they make a figure of eight – known as the waggle dance – which incorporates various kinds of semantic information.[13] The worker bee who performs the dance translates the horizontal directions into up and down. The eight consists of two semicircles and a straight line. The worker bee first makes one semicircle, returning to the starting point, then as she traces the straight line she waggles her rear end. Then she repeats the semicircle and the line back in the other direction, completing the figure of eight. The angle that the worker bee makes with the vertical axis is equal to the angle of the route to the food in relation to the sun. The distance is indicated by the speed with which the bee shakes her rear end: the faster she moves, the closer it is. The speed and the length of the dance indicate how much nectar there is – faster means more. The dancing bee also gives the other bees a sample of the scent and taste, usually before the start of the dance, so that the other bees know what to look for. In addition to dancing, bees sometimes also make sounds that give information about the distance.

They have other kinds of dances as well, including one to indicate that other bees will have to help fetch the nectar and another for starting and stopping foraging. Bees also dance to find the best location for a new nest. Deliberation is involved in this process. This is how it works: a number of scout bees go out to investigate possible nest locations and make a judgement about the various locations; they first decide if it is worth dancing about – only the best locations are – and then the length of the dance shows how good the location is. Other bees may follow to the best locations, and then dance about too. This is a collaborative process, and the very best nesting place is the one left at the end of all dancing.

Bees from different communities have their own dances, or perhaps dialects. In addition to movement, gestures and sounds, bees also make use of scent, although we have only just begun to understand the complexity of these scents. It has been argued that the composite scent signals used have their own grammar. What becomes clear when looking at the different ways bees communicate is that it can certainly be called a language: bees are able to convey abstract information by using signs.

In bee grammar, movement, sound, smell, visual signs and taste can play a role. In other species, grammar can also be found in the interplay between different physical movements. An example is found in the communication of Jacky dragons. They communicate with one another in four ways: through their posture, the number of feet they have on the ground, by nodding their heads and by puffing out their throats. This might appear simple and unsophisticated, but there are in fact 6,864 possible combinations, 172 of which are frequently used. The sequence of the actions and the length of the action are also important to their meaning, which suggests a system of grammar.[14]

The *Hylodes japi* frog, which was recently discovered in the Serra do Japi mountains in Brazil, also combines vocalisations with movements, gestures and posture. They run and jump, wave their toes and stretch their legs, raise their arms in the air and wave their hands, shake hands, twist their bodies, do silly walks, and so on. They also make head movements, drawing figures of eight with their faces, for example, and taking hold of their feet to show off their toes. Researchers have charted eighteen different types of vocalisation so far, including songs with more than five notes. Females and males also have special ways of touching each other, something that has not previously been observed in frogs, allowing them to communicate complex messages.[15]

Twenty-hour love songs

Humpback whales live mainly underwater, where sight and scent are less useful for communication. Sound, on the other hand, is very suitable as it travels faster and further through water than through the air. To human ears, whale song sounds improvised and somewhat ethereal, which is perhaps the reason why it is used for meditation. But as free flowing as these songs may sound, researchers have demonstrated that they involve grammar.[16] Humpbacks string together sounds to make sentences – using syntax – forming songs that can last up to twenty hours. The humpback researcher Ryuji Suzuki and his colleagues developed a computer program to study them,[17] breaking up all the songs into sounds and allocating them symbols. Then they used a mathematical model to analyse the patterns. They also asked humans to listen to the sounds, and by ear they reached the same conclusions as the computer.

Humpbacks combine short and long sentences into melodies, which are repeated in different keys. Songs may be long or short and can contain as few as six elements and as many as four hundred. Male humpbacks sing for six months of the year. Each season a group sings a new song, and although everyone sings the same songs the melodies become progressively complex over the course of the season, eventually becoming completely different. Different groups sing their own songs – it seems to be a matter of culture – although another group may sometimes pick up on popular songs, and it becomes a hit. These whales' songs rhyme too, often ending on the same sound. The tones and clicks that whales make when not singing also differ from one area to the next, both in sound and combination, which leads researchers to think that they are like human dialects, or even languages.[18] There are species of whale in which every individual sings their own song. Bowhead whales live in the Arctic and sing with two voices at the same time (at a high and low frequency).[19] It is not just whales that sing new songs every year. Male yellow-rumped caciques, for example, sing between five and eight songs every year, which change by seventy-eight per cent each year.[20] The songs of village indigobirds also change. This shift can sometimes take eight years, even though the birds themselves only live for eighteen months, so this is clearly a case of cultural transmission.[21]

The inaudible

Free-tailed bats make use of echolocation to navigate and to catch prey. They make sounds that are mostly too high-pitched for our ears to perceive. The echo of the sound allows them to hear their environment; the higher the sound, the more precisely this happens. They also have various other vocalisations that are not easily audible by humans. For this reason, the songs of bats were not studied for a long time, but now digital recording equipment has made such research possible. This has revealed their language to be actually very complex – bats are now believed to be the mammals with the most complex form of vocal communication after humans. Study of the songs that male free-tailed bats sing to court females shows that each male designs his own song. These songs do have fixed elements and certain patterns, but every male uses his own syllables and sounds, such as squeaks, chirps, trills and hums. The songs are constructed like human sentences. Bats use complex communication not only in love, but also in defending their territory, defining social status, bringing up their young, chasing away intruders and identifying one another.[22] They are mammals with brains similar to human brains and so in order to learn more about the origins of language, the brains of bats are now being researched.

Another animal that sings in frequencies at and above the limits of human hearing is the mouse. Female mice prefer complex to simple songs, so male mice sing complex songs to attract them. If they can only smell the female, the male's song is more complex than if she is actually present. Laboratory mice, who learn songs from one another, each have their own unique song.[23] Some songs are innate; laboratory mice who grow up with mice from other litters still sing the songs they were born with.[24] Wild mice also sing.[25] The song variations between species of mice are greater than between different birds, and the songs of mice sometimes become more complex as they get older. As mice are not valued as much as birds, their songs have not been studied in as much detail yet.

In 2015, it was discovered that females sing

back. For human ears, it is impossible to tell if mice are singing and so it was once assumed that only males sing.[26] Humans often assume that only the male of a species sings, which is based on stereotypes regarding gender and the role of language among animals: it is believed that animals sing or speak mainly in order to find a partner or to defend their territory (out of instinct, not intelligence) and that it is the males who play the active role in this. Feminist philosophers of science argue that this is based on gender prejudice.[27] In some kinds of cicada – small jumping bugs that lay eggs underground for seventeen years before all hatching at once – the males use their voices and the females make sounds with their wings. In mating conversation the male will say something with his voice and the female will respond by flapping her wings. The male repeats himself, and if the female responds again, he repeats himself once more, in a higher pitch, and if the female responds again, they mate.[28]

Like mice, some insects, such as moths[29] and grasshoppers,[30] communicate with sound that is too high for human hearing. Moths and grasshoppers have a kind of membrane in their abdominal cavities with which they perceive sound. Crickets pick up sound with their front legs[31] and mosquitoes hear with a vibration-sensitive organ at the base of their antennae.[32] Some insects hear mainly by feeling; sound makes objects move and they can feel these vibrations in the body. Sharks make water move with their body postures, which other sharks can feel and interpret. They also make use of sound, scent and electrical signals.[33] Both the vibrations of the water and this electrical communication are very difficult for humans to perceive well and study.

Embodied grammar

Bats, birds, bees and other animals have structures in their languages that can be compared to the structure of human languages. Research does not yet give us an unambiguous answer to the question of whether animal languages have grammar – this depends also on your definition of grammar – and how this can be compared to the grammar of Dutch or English. They do show, though, that it is not a strange question. The more we learn about animal communication, the more complex it appears to be, and humans are learning more and more about it.

One of the challenges in assessing the grammar of other animals is the role played by body language. Facial expressions, posture and movements might seem like primitive elements of communication, but

Wittgenstein points out in his discussion about aesthetic judgements – which he sees as complex – that it is precisely in the non-verbal judgement that the connoisseur can be recognised:[34] by a nod, a posture, the way in which someone hums their agreement, or the sound of a single word. Other animals too are able to obtain a lot of information from subtle physical clues, from the position of an ear to the angle of a tail. In order to appreciate their language, we have to learn which movements have meaning and which do not, and what those meanings are. Technological advances can help here, in analysing video recordings of movements, but also with recording utterances that we are unable to perceive, and in digitally analysing data. Sometimes, though, this still remains a challenge: the sounds made by elephants are so deep, for example, that very large speakers are needed to reproduce them and transporting these into the jungle and hiding them from the elephants is quite a feat.

It is also, again, important here to question the framework itself: if we see human grammar in the formal sense as the framework for all grammar, it becomes difficult to appreciate animal grammars. This also suggests that animal languages are inferior to human languages, as human language serves as the starting point. Wittgenstein's idea about grammar as a framework for meaningful interaction works better here, precisely because it is looser and open to other types of rules. Within human language we also find different language games, that can create meaning in different ways. Poetry can play with the rules of grammar, or question them, and yet it is still meaningful, sometimes precisely for that reason.

CHAPTER

6

METACOMMUNICATION

A dog sees another dog, runs towards him, stops abruptly a couple of metres in front of him and bows. Her front goes down while her back legs remain standing. She wags her tail. If the other dog does not respond, she barks a challenge. This position is a 'play bow'.

Marc Bekoff spent years studying the play of dogs, wolves and coyotes.[1] In play, these animals employ behaviour that normally occurs in a different situation – fighting, running away, attacking or making sexual advances – and to indicate that it is instead play, they use play signals. The play bow is the most important example of such behaviour. The exact position can vary – the front always goes down, with the rear end up high, but wagging the tail, barking, growling and making other movements is optional. Other dogs, wolves and coyotes recognise this movement and understand it as a request for play. The play bow is used at the beginning of the game, to invite the other to play, and then during play to challenge the other if he or she appears to be losing interest, and to show that it is just a game if things get a little rough – if one dog bites the other too hard or knocks them over. It can mean both 'I want to play' and 'I still want to play', with a possible 'sorry about that' thrown in.

In play, dominant dogs sometimes assume submissive positions, and vice versa. Coyotes do it too, but only with someone they know well. They also engage in self-handicapping, deliberately making themselves slower or lower in order to allow enjoyable play with a weaker or smaller coyote. Certain social conventions remain in place in dog play though: a dominant animal will rarely if ever lick another dog's mouth, and mounting the other dog is usually a one-sided affair too.[2] In play hunting – which may include running after the other, jumping on someone and tackling them – roles can be reversed. Generally in play, there is an interaction between cooperation and competition – animals test their strength and work together.

Dogs and other animals do not only play for fun. Bekoff points out that play refers to a large number of actions and expressions, which also seem to be very different in nature depending on the species. Play doesn't appear always to have a function: creativity is important, as is pretence.

Animals show that they understand the intentions of others; a fighting stance in play means something different from when it is used outside of play. Signals like the play bow establish this framework. Looking at the other's posture is also important, as are vocalisations and eye contact – animals constantly look at each other as they play. By playing, animals also learn about their own strength and position in the group and that of others.

Dogs are of course not the only animals that play, and recent years have seen an increasing amount of research into play behaviour in animals. Most mammals play, as do birds, reptiles and fish; behaviour resembling play has been observed in cephalopods, lobsters and in insects such as ants, bees and cockroaches.[3]

From play to language

The Canadian philosopher Brian Massumi argues that play is essentially creative.[4] Although animals learn by playing, play cannot be reduced to learning behaviour or to the establishment of a pecking order. It is also not only functional; beauty and joy also play a role. Individual animals develop their own playing style, which is different for everyone. They improvise and develop preferences, which can differ according to context and can change over time. A dog might play differently with one play-mate than with another, for example, and dogs can start playing differently as they get older.

Massumi argues that all behaviour has a creative component, even behaviour that is seen as instinct-ive. If rabbits always fled in exactly the same way, for example, predators would be able to anticipate their movements. While certain elements of this behaviour are always the same – such as running away – it has to be adapted to suit the specific environment, and there is room for variation and personal input. Rabbits may run to the left or the right, hiding behind a hill or standing still for a moment, for example. Rabbits need to improvise. Massumi states that instinct and expression, under-stood as the ability to vary and improvise, are not opposites but presuppose each other. Culture plays a role in this too: animals learn from others, and style and originality are important. A creative aspect is even found in the behaviour of species such as worms. Darwin studied worms extensively, viewing them as individuals who responded to stimuli in their own way and who could also learn, which requires a form of abstraction.[5] He argued that if we observed dogs exhibiting the behaviour of worms we would not hesitate to attribute certain qualities to them, such as the ability to experience pain or fear. As worms are less like us, we are sceptical; Darwin wondered if this scepticism was justified.

Language and play are connected in a variety of ways. First, linguistic utterances can be part of play, and play is a form of communication. Second, in language there is also no hard boundary between instinct and intelligence. A number of forms of expression are innate, such as facial expressions in humans or the call for the mother in ducks. Other forms of expression – the songs of zebra finches, the writing of humans – are learned. These innate forms of language do have a creative component, though, and can be further refined. Mice can start to sing in a more complex way as they get older; humans can learn in which contexts they should and should not smile. Third, metacommunication is often involved in play: communication about communication. We can speak in language about language, or write about it, as in this book. In play, dogs say something about their use of language, understood as the movements that usually occur in other contexts, for example by performing the play bow. This is necessary in order to allow the play to happen without turning into a fight.

Humour can work in the same way. Linguistic jokes are play in language, consisting of words that do not mean what they actually mean or which question their own meaning – slapstick exaggerates movements or uses them in another context. In play, animals can use certain actions outside of their context, effectively communicating about those actions with members of their own species, but also with humans. I previously discussed the example of Vicki Hearne, who taught Salty the dog to fetch a dumbbell. Salty turned this into a game, deliberately doing something that Hearne had not asked her to – she fetched the lid of the rubbish bin or took the dumbbell to someone else. Hearne saw this as a joke, which was made possible because Salty had learned the rules of a game. What Salty did is similar to the way dogs play with one another; within the context of a game actions mean something different, and you can act in an original way in order to make the game more fun. Simply taking a dumbbell to a stranger has no point, while in a game it can serve as a challenge to the other participant.

From play to morality

A game has certain rules. By playing with others, animals learn those rules. They can test themselves against other animals in a safe form of competition and work together with others through play, not a real fight. Marc Bekoff is not investigating this just for the fun of it though; his work on animal play is part of a research project into the evolution of morality. Together with Jessica Pierce, he has written a book about morality and justice in animals, based on social codes, which in part are determined and learned through play.[6] They argue that morality in humans did not evolve in a vacuum and that there is continuity between the species. If qualities such as morality, love or justice can be found in humans then it is likely that these qualities are also to be found in other species, tailored to suit.

Bekoff sees an example of the connection between morality and play in the fact that a play bow is practically never used in the wrong way. It very rarely happens that an animal indicates that he wants to play but then starts fighting instead. If this happens, the other dogs no longer want to play with the aggressor and it results in social exclusion. In play dogs set social rules, and young dogs learn about these. By assuming the role of another and by playing less roughly than they are capable of, they mark out what is socially acceptable in a group. Within the safety of the game, this is possible; a wrong move, such as a bite that is too hard, does not lead directly to fighting as a dog can indicate with a play bow that it was not meant as aggression. Play is voluntary and open, which makes it a good environment for defining boundaries. Learning to play is therefore important for young animals, for their social, cognitive and physical development. If an animal does not play with other animals at a young age, he or she will lack an awareness of group norms and values in later life.

Morality and social interaction

It is often thought that non-human animals do not – and cannot – act morally as they lack the intellectual faculties to do so. This indicates a particular conception of morality: that considered decisions are made about actions. However, recent research into the moral psychology of humans[7] shows that morality is primarily a question of habit and socialisation. Many moral decisions, such as rescuing someone who is drowning, are made automatically, in a split second, without any extensive thought. Humans are born with a certain social predisposition and develop this further in their childhood by living with other humans and adopting the norms of that community.

Other social animals have the same inclination towards social action, which is further developed by interaction within the group. We find varying degrees of morality in different species. Domesticated animals, which we know best, often do not act randomly or anarchically at all; with the right upbringing, just like humans, they conform to social norms and values in a shared human–non-human animal community.[8] This makes living together possible. The same is true of humans; if we had to weigh up every action in a social situation, it would not only take individuals too much time, it would also put the stability of society at risk.

The body plays an important role in this social conception of morality.[9] By acting with others, we add certain norms and rules to our physical repertoire, which we then express with our bodies. Sometimes we act morally without being aware of it; sometimes our body does something first and we think about it afterwards. We show who we are by acting, not only by what we say or by our opinions. The social framework that we share with others matters just as much here as our individual experience. The social framework of humans who live with other animals is also shaped by their presence and actions, and vice versa.

Animal morality

On 16 August 1996, a three-year-old boy fell into the gorilla enclosure at a zoo in Brookfield, Illinois.

He was unconscious. Binti Jua, an eight-year-old female gorilla, immediately went up to him. Zoo visitors began to scream, fearing that Binti Jua would harm the boy. But she picked him up and kept him away from other gorillas who might harm him, before gently handing him over to the staff. Her own baby was on her back the whole time.

Binti Jua's display of empathy isn't unique. In 1986, a male gorilla called Jambo also picked up a five-year-old who had fallen into his enclosure in Jersey Zoo, and handed the child over to zoo staff. Kuni, a bonobo who lives at Twycross Zoo in Leicestershire, found a starling in her enclosure that was unable to fly. She tried to make the bird move by pushing it a little, but when that did not help Kuni picked up the bird, climbed as high as possible up the tallest nearby tree and, spreading the starling's wings with her hands, threw the bird into the air. Kuni's attempt to help the bird fly didn't work, and the starling fell to the ground. Kuni then tried to throw the starling over the wall of her enclosure. When staff went to look later, the starling was gone; the bird had probably just needed a little time to recover.[10]

There is no consensus about the moral value of these actions. Frans de Waal sees the behaviour of Binti Jua as an act of empathy, other scientists doubt that and argue that it was learned behaviour;[11]

Binti Jua had been raised by humans and was used to being taken for examinations with her own baby. Jambo, however, was raised by his own mother. Kuni had not learned how to handle starlings but had often seen them flying around. There is no clear-cut answer about whether this was a case of empathy or not, but it is a meaningful question. We can approach this question in two ways: by figuring out what is going on in the animals' minds or bodies, or by investigating the meaning of the term. I will come back to the meaning of the term later, but first I would like to go more deeply into research into animal morality.

Bekoff and Pierce discuss morality with reference to three areas of research: cooperation, empathy and justice. They see a connection between social complexity and moral complexity in groups of animals. This seems logical – animals who live in complex social groups need more social rules in order to deal with one another harmoniously. The same connection appears to exist between social complexity and language; more interaction – and more complex interaction – calls for more words.

Research into animal morality is carried out both in laboratories and in the field. Different animals living in captivity show that they take the wellbeing of others into account. Rats and rhesus monkeys refuse to eat if it means another of their kind will receive a shock.[12] A male Diana monkey learned how to obtain food and then helped a female who did not understand how to do so, without any obvious benefit to himself.[13] Chimpanzees in captivity will open another chimp's cage so that he or she can also get to the food.[14] Honesty is not only perceived as important in relationships with others of the same species though. Capuchin monkeys refuse to work with researchers if they are treated unfairly.[15] In the wild, elephants have been found to comfort their friends[16] and to protect members of the group from others if they are unable to defend themselves.[17] Dolphins remain with sick dolphins, helping them for as long as possible, for example by forming a life raft around the ailing animal.[18] There are also anecdotes about dolphins helping humans and other animals. In 1983, a group of dolphins helped a group of pilot whales stranded on Tokerau Beach in New Zealand to find their way back to the sea. Five years previously, the same had happened in Whangarei Harbour. In 2004, a group of dolphins off the north coast of New Zealand formed a ring around a number of swimmers to protect them from a white shark. In the Red Sea, a group of divers became lost. Dolphins protected them from sharks and showed their human rescuers the way.[19]

Scientists differentiate between social behaviour and moral behaviour. For evolutionary biologists, taking care of one's own children is social behaviour, for instance, which may involve moral behaviour but is in itself not an expression of morality. Morality within an animal community does not imply that it also exists between animal communities. A fair amount of research has been carried out into social cohesion and morality among wolves. The fact that there are honest agreements between wolves does not mean that such agreements also exist between wolves and their prey – humans of course also often privilege their own kind. Belonging to a species probably matters less here than belonging to a community: domesticated animals learn to act morally, or to follow the rules of a community, in relation to individuals of different species.[20]

There may also be different forms of morality within different animal communities, which stem from different habits and customs. Animals often have manners, a kind of etiquette – for example, regarding who gets to eat first or how to greet others. Habits of this kind have moral significance as they express the norms of a group. This is not always only about taking the other into account; it can also be about personal gain. Humans may act morally, for example, because they want to avoid shame or want to benefit from the social advantages of a group. When we talk about animal morality, it does not mean that other animals have a morality identical to humans'. However, as with humans, norms and values play a role, as does consideration for others.

Cooperation

Cooperative behaviour can take place in many different situations – between partners, between two individuals who may or may not know each other, in large networks and in family situations.[21] Animals can also work together in ecosystems, and even, unintentionally, at cell level. Animals may cooperate for practical reasons, out of self-interest, concern for the other and maybe also because it feels good to do things together. Cooperation is not a separate category of behaviour but part of a larger network of social and helpful behaviour. In both humans and animals, cooperation does not depend solely on considered, informed decisions. The hormone oxytocin, for example, plays a role in mother–child relationships and between partners, and in combination with a number of other hormones it supports broader social cooperation among humans. We have previously seen that oxytocin also features in relationships between humans and dogs, and it affects relationships between animals too.

Biologists distinguish between various forms of cooperation. Kin selection is a form of altruism in which relatives are favoured over non-relatives. Ground squirrels have an alarm call for predators, but the animal who calls out is vulnerable, as the predator immediately knows where he or she is. Females of this species, who live with their relatives, call out much more frequently than males, who do not live near their families.

Mutualism is a form of collaboration in which two or more animals together of a different species can achieve something that they cannot achieve alone, and in which they see a direct result. This is considered the simplest kind of cooperation and it requires little thinking. Hunting in a group is an example of this behaviour. Groupers and giant moray eels hunt together, for example, making an agreement about the hunting beforehand by shaking their heads.[22]

Reciprocal altruism is another form of cooperation, which is not based on family connections. This has been researched primarily among primates. Grooming is one example of this behaviour: one animal will groom another, which takes some energy, expecting that the other animal will return the favour at some point. Animals mostly groom the animal that grooms them the most, and they are also more inclined to help those who have often groomed them. Altruism is not only found among primates. Impalas groom one another,[23] baboons sometimes swap grooming for holding another baboon's baby[24] and vampire bats feed members of their group.

It was long thought that only humans are capable of generalised reciprocity: helping others without knowing them or expecting something in return. Research into chimpanzees in captivity however shows that they repeatedly help humans without expecting anything in return – in one experiment they gave humans a stick that the humans were unable to reach[25] – and that they help other chimpanzees with no expectation of a reward, for example, by opening the other chimp's cage.[26] Rats also help strangers, and they are more inclined to do so if a stranger has previously helped them.[27] Although there have been only a few studies into this behaviour, and the animals in these studies lived in captivity, this shows that animal morality is more complex than was once thought.

As with other animal capacities, there is debate about what actually constitutes cooperation. Wolves hunt in packs. Individual animals attune their behaviour to one another, allowing them to capture and overpower larger prey than if they were to hunt alone. Some biologists think that the wolves cooperate with a common goal in mind, while others doubt that there is true cooperation involved; maybe they are simply hungry and know that this is the only way to get food. Observation does little to help here as the issue concerns the philosophical or conceptual definition of cooperation, even if further research into the use and role of communication in hunting might shed light on their intentions.

Not all forms of reciprocal altruism involve cooperation. Sometimes an animal does something for someone else and the other does something in return, but there is no actual cooperation. And not all forms of cooperation are reciprocal. A study of lions has shown that they do not always work together to chase off intruders and that the animals who do not cooperate are not punished for their lack of cooperation.[28] Furthermore, not all cooperation and not all altruism is moral behaviour. Bekoff and Pierce have discussed the altruism of slime moulds in this context. Slime moulds are organisms that were once classified as moulds and now are grouped with other one-celled organisms; we do not know exactly what they are. Some cells sacrifice themselves so that the rest of the slime moulds can continue to exist.[29] This is altruism, but it is lacking the element of what we think of as morality – as far as we can tell, slime moulds do not have the emotional and cognitive complexity that we associate with morality. So Bekoff and Pierce argue that we should only call behaviour moral in animals who live in more complex social relationships, in groups with clear standards of what is good and bad, flexibility of behaviour, and a rich emotional life. This certainly feels plausible. But because we still know little about many species, it also seems wise to be cautious when drawing lines and in measuring other animals by how closely they resemble human beings.

Empathy and emotional communication

Empathy is used in biology and ethology to indicate a class of behaviour. A simple form of empathy is emotional contagion: for example, becoming scared when someone else is scared. This can be a physical, instinctive reaction. Many animals experience this form of empathy and it has recently been demonstrated even in woodlice.[30] More complex forms are helping others, cognitive empathy – rationally understanding how the other feels – and attribution – using the imagination to assume the other's perspective. Empathy is regarded as a form of emotional communication. In this, facial expressions can play an important part. Wolves are very social animals and they have more sophisticated facial expressions than coyotes or foxes.[31]

Many different animals are known to behave empathetically. Darwin wrote about a Captain Stansbury who found an old, fat, blind pelican who was clearly being fed, and thus kept alive, by other pelicans. Crows also take care of their blind comrades, he wrote, and he had even heard a story about a blind cockerel being cared for by others of its kind.[32] There is much research currently being undertaken into empathy in rats, because their DNA is similar to that of humans. Paradoxically, this research is often very cruel as it involves testing how animals react to others' pain. Animals feel empathy not only towards others of their species, but also towards other species. Companion animals are known to empathise with their humans, sometimes attempting to comfort them. There are also stories about groups of animals taking in human children and raising them.[33]

In addition to studying behaviour, researchers have also investigated how empathy looks in the brain. They found that mirror neurons are active in the same way when an animal watches another performing an action as when the animal performs an action itself. These neurons are thought to play a role in understanding and interpreting others' actions and offer clues about their thinking, and they are probably of importance in acquiring language and emotional insight. Mirror neurons have been found in the brains of humans, other primates and birds. Spindle neurons – which are named after their long, spindle shape – are credited with allowing us to feel love and to suffer emotionally. It was long thought that these were only found in humans and other great apes and that they were what set us apart from other mammals, but spindle neurons have now been found in whales and elephants, where it is likely that they also play a role in empathy, social organisation, language and intuition about others' feelings.[34]

Research into the emotional lives of animals is in fashion at the moment, as is research into morality and language. Emotions are in biology understood as psychological phenomena that help to control behaviour. This sounds simple, but Bekoff[35] points out that the concept of emotions is actually very hard to define, perhaps because it is so general or because no single theory about the subject encompasses its complexity. What is clear, though, is that emotions exist and that they are extremely important in social relations. We can use our senses to read others' emotions, which they may show with their posture, scent, sound or expressions. Others can read our emotions in the same way. There are primary – instinctive – emotions and secondary emotions, which are consciously felt and experienced and upon which we reflect. Emotion and cognition are connected in humans and in other animals; we just do not yet know much about exactly how they are linked.

Bekoff discusses many examples of fear, joy, sadness, love, anger and even shame in non-human animals. He stresses that further research into animal emotions should be interdisciplinary and that we need to learn more about how animals live so that we can better understand why they do what they do and why they feel what they feel. Assuming that animals feel nothing or that we will never understand them is unproductive and results in research questions that will only confirm that image. Maybe dogs feel emotions in a very different way from humans, but this does not mean that there is no such thing as dog sadness or dog joy. Animal emotions can resemble those of humans in very different species. Honeybees who have just experienced an attack feel pessimistic and believe the glass is half empty, while honeybees who did not experience that attack feel optimistic and see the glass as half full.[36] Dogs[37] and elephants[38] can suffer from post-traumatic stress disorder. We have seen that Tilikum the orca became depressed with his life in captivity and having to perform tricks for humans.

Some animals feel empathy not only for their living fellow creatures, but also for their dead. Chimpanzees are known to mourn, for example, as first described by Jane Goodall, and elephants, too, have mourning rituals. Julie Ann Smith writes about mourning in rabbits. Elephants continue to visit the graves of loved ones for years and are, as previously discussed, also interested in the bones of elephants unknown to them, which suggests an understanding of death that transcends the individual's situation. Recent research into giraffes has revealed similar mourning rituals.[39] Michael the gorilla spoke in sign language about poachers killing his parents. Crows bury members of their group. This has also been observed in foxes. These are all forms of caring for others that extend beyond self-interest and everyday interaction.[40]

Some researchers see the rituals of certain animals as spiritual or even religious.[41] In humans, religion is not only something that is considered with the mind but is embedded within, and consists primarily of practices such as washing and praying. Individuals and communities are shaped by daily habits and by rituals, including religious ones. Without wishing to devalue the human spiritual experience, Jane

Goodall and Barbara Smuts have written about experiences of other animals – chimpanzees for Goodall and baboons for Smuts – that, at least to the human observer, seem spiritual. Smuts described how the baboons she was studying sat down around small pools of water and stared into them, as if very deep in thought, and then all slowly rose again at the same time and went onward in silence. She observed such a ritual twice, and noticed that even the noisy youngsters became quiet.[42] Goodall describes a chimpanzee dance by a waterfall in Gombe that appeared to have no practical purpose but instead to be inspired by feelings of awe and wonder.[43] Elephant researcher Katy Payne, who founded the Elephant Listening Project, related in a radio interview how elephants in a group sometimes become completely still, all at the same time. They stay in this position for a minute or longer, being silent. She is herself a Quaker and said this reminded her of being quiet together at Quaker meetings; she saw it as a form of meditation.[44]

Rules and justice

Darwin was already of the view that animals have a conscience and can distinguish between good and bad,[45] a theory that he based on observations, anecdotes and research. Bekoff and Pierce endorse this and argue that empathy is the foundation of moral behaviour. We can see by looking at animals that they feel things, and sometimes feel them ourselves too, because empathy can be an interspecies phenomenon. The human sense of justice is thought by some to be most highly developed, but we do not know this for certain. Orcas have very large brains with an area that humans do not have, adjacent to the limbic system, which deals with the processing of emotions. For this reason, some scientists think they are probably more social than humans and have a richer emotional life.[46] Many other animals also have a sense of justice.

The sense of fairness in chimpanzees and children has been researched with a modified version of the 'ultimatum game', in which individuals have to choose between two tokens. One option divides up the reward equally; the other favours the chooser and disadvantages the partner. The partner has to hand in the token, thereby accepting the agreed reward. It was previously thought that this game could not be played with animals because it was assumed they would always choose selfishly. This proved not to be true. Chimpanzees and children made the same choices as adults: if they need a partner's help, they share fairly; if the partner is passive, they choose selfishly.[47]

Research also shows that sensitivity to inequality is not limited to apes and primates. In a study of dogs that involved getting them to shake paws, the dogs did so, except when they saw that another dog was getting a reward for shaking paws and they themselves were not. They also exhibited more symptoms of stress. The researchers saw this as a sign that dogs have a sense of what is fair and what is not.[48] Dogs also judge the social behaviour of humans with one another, demonstrating an empathetic loyalty to their own human. A recent study required the dogs to watch people helping their humans to open a box, and also people who did not help their humans. If the non-helpers offered the dog a snack afterwards, the dog usually refused it. They would, however, accept a biscuit from the helpers.[49]

The research of emotions and morality also invites us to take a closer look at human morality. Humans tend to think that they are particularly well developed in terms of justice and yet they are also the species that exploits and uses other animals on a huge scale for their own gain. In a broader sense, the world we live in is largely determined by human actions, which has prompted thinkers to name our current age the Anthropocene. Humans have occupied the territories of many groups of non-human animals, and, in many places, dominate shared spaces with roads, buildings, ships, but also noise and other forms of pollution. It is beyond the scope of this book to determine human obligations and responsibilities towards other animals. I do want to suggest that language research can, however, play a part in these considerations, because language gives us an insight into the inner lives of others, and can play a key role in bringing about new relationships with animals.

CHAPTER

7

WHY WE NEED TO TALK
WITH THE ANIMALS

Bats sing songs for the ones they love, which are as complex in structure as human sentences. Parrots can talk in human language to humans about mathematical problems. Dogs understand the grammar of human language and communicate with scent patterns, which have their own grammar. Bees symbolically communicate spatial coordinates through dance. Dolphins have names. Prairie dogs describe visitors in detail. Dogs and their humans produce the cuddle hormone when they have eye contact. While playing, wolves communicate about the game. Horses can read humans' bodies. Cephalopods can convey a wide range of information to others with the colour patterns on their skins. In these and other expressions of language, animals give us and one another information about how they feel and what they want, and they describe their surroundings. They make contact, ask questions and give answers. Human language is perhaps unique in its complexity and versatility, but so are the languages of other animals.

It is too early to reach final conclusions about animal languages and to give a full definition of animal language. This is because scientific research into the subject is very recent, and because humans in isolation should never reach such a conclusion. From a political point of view, it is problematic to determine for others what constitutes meaningful communication. Instead of defining whether non-human animal forms of communication fit into the frame of what humans define as 'language', we should instead pay attention to what they are saying, and begin investigating what language is and could be from there. This is not just a matter of listening: we also need to do our best to find new ways to interact with other animals about common issues. In this chapter I will investigate the role of language in forming new relationships with animals, supported by the latest literature about the role and place of animals in ethics and political philosophy.

Political animals

The view that only humans are political actors has a long history. In Book I of *The Politics*, Aristotle defines man as a political animal and the only animal that is endowed with speech, and more specifically the ability to distinguish between right and wrong. He saw this capacity as necessary for being part of the political community, and attributing it only to humans, he draws a line between humans and other animals. This line functions as a border around the political; only humans can be political animals. The notion that only humans are political actors is still widespread, not only in philosophy but also in political practice.

Like the idea that animals have no language, in recent years this premise has been challenged in different fields. In political philosophy, it has been proposed that animals should be considered as political actors who stand in different relationships to human political communities.[1] Schools of thought such as poststructuralism[2] and posthumanism[3] question human exceptionalism. This questioning has also occurred within the disciplines of animal studies, and in animal geography, a discipline that also highlights the various ways in which animals already influence human political communities.[4]

When it comes to animals and politics, humans often joke that other animals are unable to vote. Research into animal group decisions, however, shows that they can and do vote within their societies. In bee communities we can observe a process that can be seen as an animal variant of what the German philosopher Habermas called deliberation: different individuals discuss the options and collectively choose the best one.[5] Red deer start moving when more than around sixty-two per cent of the adults stand up.[6] African buffalos also make group decisions about when to stand and where to go: the females decide what will happen by standing up, looking in a particular direction and then lying down again. If there are different preferences, the group sometimes splits into smaller groups. Researchers once believed that the buffalos simply stood up to stretch their legs, but it is in fact about decision-making.[7] In groups of pigeons, the hierarchy is flexible. Different individuals are highest in rank at different times and decide where the birds will fly.[8] Cockroaches probably have less advanced forms of decision-making than bees and ants but they do not act chaotically or without logic. In an experiment with fifty cockroaches and three hiding places, the group split in two and occupied two of the three places. When the hiding places were bigger, they all went into one of them. This shows that they are looking for the right balance between cooperation and competition.[9] Dominant male and female baboons take the decisions, although other baboons influence those decisions; every action counts.[10]

I wrote earlier that concepts such as pain, fear and love do not arise in a vacuum: other animals influence them by their actions and their presence. Some thinkers argue that this also applies to political communities and to what we mean by politics. We often believe that politics is beyond animals, that they are unable to understand it. We also like to think that we as humans are alone responsible for social and societal structures. I previously discussed the work of the historian Jason Hribal, who describes how animal resistance influences human practices and social structures. When thinking about animals and politics it is important to note that politics does not only take place in the town council or parliament. There are many different counter-practices that are seen as political, in both humans and other animals, that influence official forms of political decision-making.

Human political concepts can however guide us in thinking about establishing new political relations with other animals. The political philosophers Sue Donaldson and Will Kymlicka argue that groups of non-human animals should be seen as political communities.[11] In determining rights and duties, we should look at how these communities relate to human political communities. They propose to divide non-human animals into three groups. Wild animals, those who prefer to stay away from humans as far as possible, should be considered as sovereign self-governing communities. Domesticated animals, such as companions and farmed animals, should be granted the right to citizenship. Animals who live among humans but are not domesticated should be regarded as 'denizens' and have rights of residence, but not the full set of citizenship rights.

In determining the precise content of these rights, context and animal agency matter. Domesticated animals have a right to be part of our communities, Donaldson and Kymlicka write, because historically humans have mistreated them: for example, by capturing them and by using breeding programmes to alter their bodies in such a way that they have become dependent on humans. They are now part of shared interspecies communities, this is their home, and removing them would be unjust. These animals can also be part of shared human–animal communities because they have certain characteristics that make coexistence possible – we see this for example in the shared language games discussed above. Rights for domesticated animals would include a right to healthcare, a roof over their heads and political representation. Wild animals fall on the other side of the spectrum. They usually avoid contact with humans and are able to take care of themselves. This does not mean that intervention should always be avoided: sometimes we have a duty to help others in need even though they belong to different communities, and we must also take into account the effects of our actions on their living environments. The liminal animals who live among humans in cities and rural areas will generally avoid contact with humans, but they also have rights, such as the right to a place to settle and not to be discriminated against.

The roots of this political theory are in animal rights philosophy, a branch of ethics in which other animals are seen as subjects with lives that are important to them. In thinking about animal rights, the emphasis was long placed on 'negative rights': rights that determine what should not happen to an individual – for example, the right not to be killed, not to be captured and not to be used for others' profit.[12] Donaldson and Kymlicka point out that these rights are very important to animals – human and non-human – and are necessary to change their position in society, but that they are not enough. In order to be able to lead a valuable life, it is not sufficient not to be killed or captured. You must also be able to live somewhere, to have the opportunity to enter into relationships with others and to be able to develop your skills and talents in other ways.[13] Humans and other animals already live together in many different ways, sometimes by forming communities and sometimes because they share a territory – and we share a planet. Relationships cannot be avoided, and do not have to be avoided, because better relations are possible. In order to think about dealing fairly with other animals it can be helpful to look at how human political communities relate to one another, and what we think fair in these cases.

The role of language in political interaction

Language plays an important role in political interaction both in formal situations – as in parliament – and in many other practices, from demonstrations to political campaign materials and websites. One of the characteristics of a democracy is that those who live in it are not merely able to participate in the system as it is; they also have a say in determining that system. We do not only have a passive right to vote but can also put ourselves up for election and propose new laws and regulations.

Animals we live with – dogs, cats, rabbits, guinea pigs, cows, horses, donkeys, goats, sheep, chickens and so on – have their own ideas about the good life and their own ways of communicating those to humans. It is often thought that humans may be able to take animals' interests into consideration, but that animals themselves will never be capable of helping to govern or to think about laws and the like. While animals cannot stand for election or have a meaningful discussion in parliament within the current system, this does not mean that participation is impossible and certainly not that it is undesirable.

Many laws and regulations affect the lives of other animals, even though they are not consulted. The reason often given for this is that they cannot speak. As we learn more about animal languages and subjectivity, it has become clearer that this is simply not true, and we should not ignore them any longer. Insights from civil rights movements about discrimination and equality can also be helpful in understanding what is at stake here. Similar to other dominant groups, humans should rethink their own position in society. This is not just important for the animals. Environmental and climate problems have alerted humans to the consequences of their lifestyle for future generations.

Political communication with animals already takes place – in border conflicts, households, cities and states. In order to improve this communication, species-specific languages must be taken into account. There is little point in telling a goose in human language that she is not welcome somewhere, but this does not mean communication is impossible. We can operate on the basis of their languages, and shared interspecies language games, to explain things to them. Curious animals can work together to arrive at new forms of language too, as has been shown by Lorenz's research into, and with, geese. As a consequence, we need to redefine what we understand by political communication, and by politics. Various political philosophers have criticised the image of language usage in politics as solely rational and argumentative, and have drawn our attention to the fact that it is problematic that only those who express themselves in this way are taken seriously, because other voices are not heard. They point, for example, to the importance of rituals, body language, greetings, and the role of emotions, stories and rhetoric.[14] Human communities furthermore express themselves politically in different ways, and the dominant image of political action already excludes people who come from groups that were traditionally not taken seriously in politics.[15] When a woman raises her voice in parliament, she is often described as emotional, whereas a man is seen as strong and passionate, for example.

But what exact form might political communication with animals take? Of course, this depends just as much on the animals in question as on the humans. There is a huge variety of ways in which individuals, species and communities express themselves, which cannot be captured in a few sentences – or even in a book.[16] New research into living with animals can be used to experiment in this area. For such research to succeed, it is important to assume that animals communicate meaningfully, and to keep an open mind about the future. In connection with this, we can consider case studies in which humans and other animals already communicate politically, in order to investigate how new relations can be formed, when we begin viewing other animals and their actions differently.

Political communication with wild animals: the macaques of Singapore

In the Bukit Timah nature reserve in Singapore, the population of macaques is threatened.[17] The people who have come to live in the area have blocked the macaques' access to green corridors with their houses. These humans knew that the monkeys lived there before they moved in, and they actually say that living closer to nature was their reason for choosing that area. They also feed the animals; as a result, the monkeys have become too bold and come too close to the humans, causing problems by stealing food and making noise. The humans regularly have encounters with them that they describe as annoying or frightening. The humans' attitudes towards the monkeys is not solely negative, however; many also think they are cute and believe that they should not simply be killed. When conflicts occur, the park managers weigh up the wishes of the human residents against the need to protect the macaques, with the macaques usually coming off worst; they are usually killed when they are seen as a nuisance.

One solution to the conflict could be for the humans to leave; after all, they have occupied the animals' territory and there are other places they could live. If the humans have been living there for some time, if children were born there, or if they have nowhere else to go, or if animals are entering the human territory, the situation changes and ways to coexist have to be found. Animal geographers Jun-Han Yeo and Harvey Neo, who studied this situation, listed various forms of communication between the humans and the macaques, such as eye contact, keeping distance, reading each others' body language and trying to approach the other. The macaques react to humans talking, they are sensitive to their intonation, and humans react to the noises made by the macaques. A resident called Cindy remarked, for example: 'I once reprimanded a monkey for attempting to snatch my bag from me. It seemed to understand my reaction, like raising my voice, pointing a finger at it, and it backed off.'[18] Reflecting on the form of this interaction, both by learning about the language of the monkeys and by introducing political rituals, starting with greetings, could help to create a model in which monkeys and humans all have a say, and therefore come closer to one another or define the boundaries more clearly. Macaques are already directly exerting political agency here: they are questioning the hierarchy, the property of land, and communicating with humans. The solutions that Yeo and Neo put forward are mainly about increasing knowledge on the human side, for example, by putting up signs with a warning that feeding the animals makes them aggressive. Mutual learning about one another's forms of communication could be a useful addition to this approach.

Political communication with dogs

In Los Angeles, dogs and humans worked together to make the Laurel Canyon Dog Park a safe place.[19] The area had become run down and there were problems with criminals. A group of people decided to renovate the park and illegally let their dogs roam free. As a result, the unwanted criminal visitors moved elsewhere. The park became safer and other local residents started using it again. These other residents then objected to the presence of dogs not on the leash. The renovation group succeeded in keeping the place as an off-leash area. There was linguistic interaction on all levels: between the renovation group and the humans who were causing a nuisance, the dogs and those humans, the dogs and the other local residents who used the park, and the dogs among themselves. The dogs and their humans have allowed the park to function as a meeting place; conversations can now occur on a constant basis. Although it was not the dogs themselves who came up with the idea, they were necessary for the success of the action, and they influenced the form of the interaction. The park is now a place where humans and animals like to go and where various parties take care of the park's maintenance.

Dogs can also act politically without humans

being involved. Moscow has a small pack of stray dogs living in its suburbs who regularly take the underground train to the centre to eat. The biologist Andrei Poyarkov has been studying the city's stray dogs for thirty years and refers to those who take the train as the 'intellectual elite'.[20] These dogs also know when they can cross the road, they can read the traffic lights, and have figured out which humans to ask for food – they are particularly good at reading body language, also taking style of clothing into account – mostly women over forty. Both commuters and tourists appreciate the dogs' presence in the underground. Dogs are officially not allowed to travel on the Moscow metro, but commuters sometimes let in the strays, although more often they slip through the gates themselves when it is busy. By doing so they are questioning the fact that the metro is reserved for humans and are seizing the right to travel by underground train – maybe not in an intentional human way, but with their physical presence. Their behaviour also influences existing stereotypes about stray dogs; they are, in fact, clever and skilful.[21] The dogs behave well, sitting quietly on or under a seat.

Every once in a while, the Moscow city council comes up with plans to remove stray dogs from the city, which used to mean killing them. Various groups then speak up to protect the animals, such as animal rights activists and humans who look after the dogs. The metro dogs have begun to play a role in this process, with their presence and because people are sharing pictures of them on the Internet, making them ambassadors for their species.[22] The dogs demonstrate that it is impossible to keep stray dogs off the metro and out of the city, by occupying a specifically human space. They also show that there is no need to keep them out, as they behave well. In 2001, some time after the first pictures of the metro dogs emerged on the Internet, it was made illegal to shoot street dogs.

Thinking with animals

In philosophy, there is a lot of thinking *about* animals, but not much thinking *with* animals. Thinking with animals might seem utopian or vaguely spiritual, but this does not have to be the case. Language gives us insight into what others think and a way to show them what we think. As Heidegger wrote, language gives insight into the world around us and shapes that world. Thinking and speaking with animals also has these two aspects: it teaches humans to understand them better and it provides footholds for new relationships.

Within philosophy, dialogue has long been a tried and tested means for seeking the truth. For some time now, many philosophers have no longer believed in one universal truth. There are better and worse arguments though. By entering into dialogue with one another, by trying to convince the other, by adjusting our own attitude where necessary and reformulating it, we can make better judgements and maybe achieve a better understanding of the world and our own position in it. This does not mean we will ever entirely reach the truth or ultimate knowledge – after all, we are always situated in and bound to a body, a history and a place in the world.

To find out what other animals want, it is not enough merely to study them. We need to talk with them. Having conversations with animals requires us to challenge the hierarchy between humans and other animals, but this change can also come about through dialogue, when humans begin to see animals differently. Talking with animals also requires a new way of thinking about language. Other animals show us that language is broader and richer than we thought, and that there are many more ways of expressing ourselves meaningfully than in human words alone. Rather than dismissing these forms of expression as inferior, we can learn from them – about other animals and their inner lives, and about the different ways in which meaning comes about. For animal languages to be language, other animals do not have to learn anything new; humans just need to begin to see them differently.

They have been speaking all along.

ACKNOWLEDGEMENTS

With thanks to Yolande Jansen and Miriam Reeders for reading and commenting on the first version of the manuscript. Thanks to Gerrit Meijer for years of cutting out newspaper articles about animal research. Thanks to Ruth Scherpenhuisen for supporting my love of animals when I was small. Above all, many thanks to Putih, Olli, Pika, Joy, Dotje, Punkie, Kitty, Ronja, Destiny, Poemelie, Witje I and II, Hondje, Rakker, Pino, Luna, Mickey, Muis, Pol, Saartje, Jezus and Aalmoes, Bobo, Noesja and all the others who have patiently taught me what they meant and have been my friends.

NOTES

In this book I challenge the hierarchy between humans and other animals with regard to language. In line with this, I try to not repeat stereotypical views of other animals in the words I use to describe them. Humans, for example, are commonly described as 'he', 'she' and 'they', while animals are 'it'. Humans are owners; their companions are pets. I avoid using 'it' when referring to animals, and use 'he' or 'she' if the sex is known, or 'they' and 'their' if it is not.

Introduction

1   I will return to the examples in this paragraph and discuss them in greater depth, with the exception of the mantis shrimp and the marmoset.

2   Thoen, Hanne H. et al. 'A different form of color vision in mantis shrimp', *Science* 343.6169, 2014, pp. 411–13.

3   Albuquerque, Natalia et al. 'Dogs recognize dog and human emotions', *Biology Letters* 12.1, 2016, https://doi.org/10.1098/rsbl.2015.0883.

4   Takahashi, Daniel Y., Narayanan, Darshana Z. and Ghazanfar, Asif A. 'Coupled oscillator dynamics of vocal turn-taking in monkeys', *Current Biology* 23.21, 2013, pp. 2162–8.

5 Allen, Colin and Bekoff, Marc. *Species of Mind: The Philosophy and Biology of Cognitive Ethology*, MIT Press, 1999.

6 For the intersections of sexism and speciesism, see for example Adams, Carol J., *The Sexual Politics of Meat: A Feminist-Vegetarian Critical Theory*, A&C Black, 2010.

7 Wittgenstein, Ludwig. *Filosofische onderzoekingen*, Uitgeverij Boom, 2006.

8 Derrida, Jacques and Mallet, Marie-Louise. *The Animal That Therefore I Am*, Fordham University Press, 2008.

9 Kleczkowska, Katarzyna. 'Those who cannot speak: animals as others in ancient Greek thought', *Maska* 24, 2014, pp. 97–108.

10 As I mentioned earlier, I am discussing animals in the Western philosophical tradition. See Abram, David, *The Spell of the Sensuous: Perception and Language in a More-than-Human World*, Vintage, 1997, for a discussion of the role of language in other cultures, and the consequences of this for the relationship with non-human animals.

11 See Dunayer, Joan, *Animal Equality: Language and Liberation*, Ryce, Derwood MD, 2001, for a comprehensive discussion of linguistic discrimination against animals.

12 Waal, Frans de. 'Anthropomorphism and anthropodenial: consistency in our thinking about humans and other animals', *Philosophical Topics* 27.1, 1999, pp. 255–80.

13 Aristoteles, *Politica*, Historische Uitgeverij, 2012.

14 See Descartes' letter to the Marquess of Newcastle,

23 November 1646, in Descartes, René et al., *The Philosophical Writings of Descartes: Volume 3, The Correspondence*, Cambridge University Press, 1991.

15 Kant, Immanuel. *Grondslagen van de ethiek*, Boom, Amsterdam/Meppel, 1978.

16 Heidegger, Martin. *The Fundamental Concepts of Metaphysics: World, Finitude, Solitude*, Indiana University Press, 2001.

17 See Descartes' letter to the Marquess of Newcastle, 23 November 1646, op. cit.

Chapter 1: Speaking in Human Language

1 http://www.nrc.nl/ik/2015/01/26/hoest/

2 Pepperberg, Irene M. *The Alex Studies: Cognitive and Communicative Abilities of Grey Parrots*, Harvard University Press, 2009.

3 Burger, Joanna. *The Parrot Who Owns Me: The Story of a Relationship*, Villard, 2001.

4 Lorenz, Konrad and Kerr Wilson, Marjorie. *King Solomon's Ring: New Light on Animal Ways*, Psychology Press, 2002.

5 Chartrand, Tanya L. and Baaren, Rick B. van. 'Human mimicry', *Advances in Experimental Social Psychology* 41, 2009, pp. 219–74.

6 Baaren, Rick B. van et al. 'Mimicry and prosocial behavior', *Psychological Science* 15.1, 2004, pp. 71–4.

7 Iacoboni, Marco. 'Imitation, empathy, and mirror neurons', *Annual Review of Psychology* 60, 2009, pp. 653–70.

8 Kellogg, W. N. and Kellogg, L. A. *The Ape and the*

Child, Anthropoid Experiment Station of Yale
University, 1932.

9   Hayes, Keith J. and Hayes, Catherine. 'Imitation in
a home-raised chimpanzee', *Journal of Comparative
and Physiological Psychology* 45.5, 1952, pp.
450–9.

10 Gardner, Allen and Gardner, Beatrix. *Teaching Sign
Language to the Chimpanzee Washoe*, Penn State
University, Psychological Cinema Register, 1973.

11 Hess, Elizabeth. *Nim Chimpsky: The Chimp Who
Would Be Human*, Bantam, 2008.

12 Savage-Rumbaugh, E. Sue, Rumbaugh, Duane M. and
Boysen, Sarah. 'Do apes use language? One research
group considers the evidence for representational
ability in apes', *American Scientist*, 1980, pp. 49–61.

13 Patterson, Francine G. 'The gestures of a gorilla:
language acquisition in another pongid', *Brain and
Language* 5.1, 1978, pp. 72–97.

14 http://www.koko.org/michaels-story

15 Patterson, F. and Gordon, W. 'Twenty-seven years of
Project Koko and Michael', *All Apes Great and Small*
1, 2002, pp. 165–76.

16 Savage-Rumbaugh, Sue, Shanker, Stuart G. and Taylor,
Talbot J. *Apes, Language, and the Human Mind*,
Oxford University Press, 1998.

17 Hearne, Vicki. *Adam's Task: Calling Animals by
Name*, Skyhorse Publishing Inc., 1986.

18 Nishimura, Takeshi et al. 'Descent of the larynx in
chimpanzee infants', *Proceedings of the National
Academy of Sciences* 100.12, 2003, pp. 6930–3.

19 Hobaiter, Catherine and Byrne, Richard W. 'The

meanings of chimpanzee gestures', *Current Biology* 24.14, 2014, pp. 1596–1600.

20 Roberts, Anna Ilona et al. 'Chimpanzees modify intentional gestures to coordinate a search for hidden food', *Nature Communications* 5, 2014.

21 Leeuwen, Edwin J. C. van, Cronin, Katherine A. and Haun, Daniel B. M. 'A group-specific arbitrary tradition in chimpanzees (Pan troglodytes)', *Animal Cognition* 17.6, 2014, pp. 1421–5.

22 Lilly, John Cunningham. *Man and Dolphin*, Doubleday, 1961.

23 Little research has been done into suicide among animals. The ethologist Marc Bekoff (2012) writes in a blog that there is anecdotal evidence of animals ending their own lives when they are deeply unhappy. He refers to elephants standing on their trunks or stepping off cliffs, whales intentionally beaching themselves and cats jumping from heights after earthquakes. He also discusses the case of a donkey who lost her baby and walked into the water to drown herself. See: https://www.psychologytoday.com/blog/animal-emotions/201207/did-female-burro-commit-suicide

24 See the BBC documentary *The Girl Who Talked to Dolphins*.

25 Herzing, Denise L. *Dolphin Diaries: My 25 Years with Spotted Dolphins in the Bahamas*, Macmillan, 2011.

26 Ridgway, Sam et al. 'Spontaneous human speech mimicry by a cetacean', *Current Biology* 22.20, 2012, https://doi.org/10.1016/j.cub.2012.08.044.

27 Pogrebnoj-Alexandroff, A. *The True History or Who is Talking? An Elephant!*, Lode Star Publishing, 1993.

28   Stoeger, Angela S. et al. 'An Asian elephant imitates human speech', *Current Biology* 22.22, 2012, pp. 2144–8.

29   In inaccessible mountain areas, some humans use a whistled language to communicate over long distances. One example of such a language is Silbo Gomero, which is spoken by some inhabitants of La Gomera in the Canary Islands.

30   For more information about elephants, see the Elephant Listening Project website: http://www.birds.cornell.edu/brp/elephant/

31   O'Connell, Caitlin. *Elephant Don: The Politics of a Pachyderm Posse*, University of Chicago Press, 2015.

32   Bradshaw, Isabel Gay A. 'Not by bread alone: symbolic loss, trauma, and recovery in elephant communities', *Society & Animals* 12.2, 2004, pp. 143–58.

33   Lorenz, Konrad and Kerr Wilson, Marjorie. *King Solomon's Ring: New Light on Animal Ways*, Psychology Press, 2002.

34   Westerfield, Michael. *The Language of Crows*, Ashford Press, 2012.

35   Ibid.

36   St Clair, James J. H. et al. 'Experimental resource pulses influence social-network dynamics and the potential for information flow in tool-using crows', *Nature Communications* 6, 2015; http://phys.org/news/2015-11-crows.html

37   Marzluff, John M. et al. 'Lasting recognition of threatening people by wild American crows', *Animal Behaviour* 79.3, 2010, pp. 699–707.

38   Healy, Susan D. and Krebs, John R. 'Food storing
      and the hippocampus in corvids: amount and volume
      are correlated', *Proceedings of the Royal Society of
      London B: Biological Sciences* 248.1323, 1992,
      pp. 241–5.

39   Pika, Simone and Bugnyar, Thomas. 'The use of
      referential gestures in ravens (Corvus corax) in the
      wild', *Nature Communications* 2, 2011.

40   Taylor, Alex H. et al. 'Complex cognition and
      behavioural innovation in New Caledonian crows',
      *Proceedings of the Royal Society of London B:
      Biological Sciences*, 277.1694, 2010, https://doi.org/
      10.1098/rspb.2010.0285; https://www.wimp.com/
      a-crow-solves-an-eight-step-puzzle.

41   Swift, Kaeli. *Wild American Crows Use Funerals to
      Learn about Danger*, Diss., University of Washington,
      2015.

42   Wittgenstein, Ludwig. *Filosofische onderzoekingen*,
      Uitgeverij Boom, 2006.

43   Gaita, Raimond. *The Philosopher's Dog: Friendships
      with Animals*, Random House, 2009.

44   Hare, Brian and Woods, Vanessa. *The Genius of
      Dogs: Discovering the Unique Intelligence of Man's
      Best Friend*, Oneworld Publications, 2013.

45   http://www.bbc.com/earth/story/20150216-can-any-
      animals-talk-like-humans

46   Musser, Whitney B. et al. 'Differences in acoustic
      features of vocalizations produced by killer whales
      cross-socialized with bottlenose dolphins', *Journal of
      the Acoustical Society of America* 136.4, 2014, pp.
      1990–2002.

47   Lameira, Adriano R. et al. 'Speech-like rhythm in a

voiced and voiceless orangutan call', *PLoS One* 10.1, 2015, https://doi.org/10.1371/journal.pone.0116136.

48 Murayama, Tsukasa et al. 'Preliminary study of object labeling using sound production in a beluga', *International Journal of Comparative Psychology* 25.3, 2012, pp. 195–207.

49 Slobodchikoff, Con. *Chasing Doctor Dolittle: Learning the Language of Animals*, Macmillan, 2012.

Chapter 2: Conversations in the Living World

1 For more detailed information about the language of prairie dogs, see Slobodchikoff, Constantine Nicholas, Perla, Bianca S. and Verdolin, Jennifer L., *Prairie Dogs: Communication and Community in an Animal Society*, Harvard University Press, 2009.

2 For more information about the languages of the chickadee and the chicken, see Slobodchikoff, Con, *Chasing Doctor Dolittle: Learning the Language of Animals*, Macmillan, 2012.

3 Seyfarth, Robert M., Cheney, Dorothy L. and Marler, Peter. 'Vervet monkey alarm calls: semantic communication in a free-ranging primate', *Animal Behaviour* 28.4, 1980, pp. 1070–94.

4 Zuberbühler, Klaus. 'A syntactic rule in forest monkey communication', *Animal Behaviour* 63.2, 2002, pp. 293–9.

5 Flower, Tom. 'Fork-tailed drongos use deceptive mimicked alarm calls to steal food', *Proceedings of the Royal Society of London B: Biological Sciences* 278.1711, 2011, pp. 1548–55.

6   Breure, Abraham S. H. 'The sound of a snail: two
     cases of acoustic defence in gastropods', *Journal
     of Molluscan Studies* 81.2, 2015, pp. 290–3.

7   Boch, R. and Rothenbuhler, Walter C. 'Defensive
     behaviour and production of alarm pheromone in
     honeybees', *Journal of Apicultural Research* 13.4,
     1974, pp. 217–21.

8   Vander Meer, Robert K. et al. *Pheromone
     Communication in Social Insects: Ants, Wasps, Bees
     and Termites*, Westview Press, 1998.

9   De Bruijn, P. J. A. *Context-Dependent Chemical
     Communication, Alarm Pheromones of Thrips
     Larvae*, PhD thesis, University of Amsterdam, 2015.

10   Gibson Hill, C. A. 'Display and posturing in the cape
     gannet, Morus capensis', *Ibis* 90.4, 1948, pp. 568–72.

11   Fry, C. Hilary and Fry, Kathie. *Kingfishers, Bee-eaters
     and Rollers*, A&C Black, 2010.

12   Clayton, Nicola S., Dally, Joanna M. and Emery, Nathan
     J. 'Social cognition by food-caching corvids: the western
     scrub-jay as a natural psychologist', *Philosophical Trans-
     actions of the Royal Society of London B: Biological
     Sciences* 362.1480, 2007, pp. 507–22.

13   For more information about canine communications
     and cognition, see Hare, Brian and Woods, Vanessa,
     *The Genius of Dogs: Discovering the Unique
     Intelligence of Man's Best Friend*, Oneworld
     Publications, 2013.

14   Slobodchikoff, Con. *Chasing Doctor Dolittle*, op. cit.

15   Smuts, Barbara B. and Watanabe, John M. 'Social
     relationships and ritualized greetings in adult male
     baboons (Papio cynocephalus anubis)', *International
     Journal of Primatology* 11.2, 1990, pp. 147–72.

16   Smuts, Barbara. 'Gestural communication in olive
     baboons and domestic dogs' in Bekoff, Marc, Allen,
     Colin and Burghardt, Gordon M. (eds), *The
     Cognitive Animal: Empirical and Theoretical
     Perspectives on Animal Cognition*, MIT Press, 2002,
     pp. 301–6.

17   Allen, Colin and Bekoff, Marc. *Species of Mind: The
     Philosophy and Biology of Cognitive Ethology*, MIT
     Press, 1999.

18   Barton, Robert A. 'Animal communication: do
     dolphins have names?', *Current Biology* 16.15, 2006,
     https://doi.org/10.1016/j.cub.2006.07.002.

19   Burger, Joanna. *The Parrot Who Owns Me: The Story
     of a Relationship*, Villard, 2001.

20   Newman, John D. 'Squirrel monkey communication'
     in *Handbook of Squirrel Monkey Research*, Springer
     US, 1985, pp. 99–126.

21   Smith, Richard L. 'Acoustic signatures of birds, bats,
     bells, and bearings', Annual Vibration Institute
     Meeting, Dearborn, MI, 1998.

22   Burgener, Nicole et al. 'Do spotted hyena scent marks
     code for clan membership?', *Chemical Signals in
     Vertebrates* 11, 2008, pp. 169–77.

23   Bekoff, Marc. 'Observations of scent-marking and
     discriminating self from others by a domestic dog
     (Canis familiaris): tales of displaced yellow snow',
     Behavioural Processes 55.2, 2001, pp. 75–9.

24   Slobodchikoff, Con. *Chasing Doctor Dolittle*, op. cit.

25   Corson, Trevor. *The Secret Life of Lobsters: How
     Fishermen and Scientists Are Unraveling the Mysteries
     of Our Favorite Crustacean*, HarperCollins, 2004.

26   Scott, Mitchell L. et al. 'Chemosensory discrimination

of social cues mediates space use in snakes, Cryptophis nigrescens (Elapidae)', *Animal Behaviour* 85.6, 2013, pp. 1493–1500.

27 Miller, Ashadee Kay et al. 'An ambusher's arsenal: chemical crypsis in the puff adder (Bitis arietans)', *Proceedings of the Royal Society of London B.* 282.1821, 2015, https://doi.org/10.1098/rspb.2015.2182.

28 Young, Bruce A., Mathevon, Nicolas and Tang, Yezhong. 'Reptile auditory neuroethology: what do reptiles do with their hearing?', *Insights from Comparative Hearing Research*, 2014, pp. 323–46.

29 Palacios, V. et al. 'Recognition of familiarity on the basis of howls: a playback experiment in a captive group of wolves', *Behaviour* 152.5, 2015, pp. 593–614.

30 Hansen, Sara J. K. et al. 'Pairing call response surveys and distance sampling for a mammalian carnivore', *Journal of Wildlife Management* 79.4, 2015, pp. 662–71.

31 Déaux, Éloïse C. and Clarke, Jennifer A. 'Dingo (Canis lupus dingo) acoustic repertoire: form and contexts', *Behaviour* 150.1, 2013, pp. 75–101.

32 Salinas-Melgoza, Alejandro and Wright, Timothy F. 'Evidence for vocal learning and limited dispersal as dual mechanisms for dialect maintenance in a parrot', *PLoS One*, 2012, https://doi.org/10.1371/journal.pone.0048667.

33 Slobodchikoff, Con. *Chasing Doctor Dolittle*, op. cit.

34 Aplin, Lucy M. et al. 'Experimentally induced innovations lead to persistent culture via conformity in wild birds', *Nature* 518.7540, 2015, pp. 538–41.

35 Plotnik, Joshua M., Waal, Frans B. M. de and Reiss,

Diana. 'Self-recognition in an Asian elephant', *Proceedings of the National Academy of Sciences* 103.45, 2006, https://doi.org/10.1073/pnas.0608062103.

36 Shillito, Daniel J., Gallup, Gordon G. and Beck, Benjamin. 'Factors affecting mirror behaviour in western lowland gorillas, Gorilla gorilla', *Animal Behaviour* 57.5, 1999, pp. 999–1004.

37 Swartz, K. B. and Evans, S. 'Social and cognitive factors in chimpanzee and gorilla mirror behaviour and self-recognition' in Parker, S. T., Mitchell, R. W. and Boccia, M. L. (eds), *Self-awareness in Animals and Humans: Developmental Perspectives*, Cambridge University Press, 1994, pp. 189–206.

38 Broesch, T. et al. 'Cultural variations in children's mirror self-recognition', *Journal of Cross-Cultural Psychology* 42.6, 2011, pp. 1018–29.

39 Bekoff, Marc. 'Observations of scent-marking', op. cit.

40 Bruckstein, Alfred M. 'Why the ant trails look so straight and nice', *Mathematical Intelligencer* 15.2, 1993, pp. 59–62.

41 Jarau, Stefan. 'Chemical communication during food exploitation in stingless bees' in Jarau, Stefan and Hrncir, Michael (eds), *Food Exploitation by Social Insects: Ecological, Behavioral, and Theoretical Approaches*, CRC Press, 2009, pp. 223–49.

42 Wilkinson, Gerald S. 'Reciprocal food sharing in the vampire bat', *Nature* 308.5955, 1984, pp. 181–4.

43 Kunz, T. H. et al. 'Allomaternal care: helper-assisted birth in the Rodrigues fruit bat, Pteropus rodricensis (Chiroptera: Pteropodidae)', *Journal of Zoology* 232.4, 1994, pp. 691–700.

44   Normand, Emmanuelle, Dagui Ban, Simone and
     Boesch, Christophe. 'Forest chimpanzees (Pan
     troglodytes verus) remember the location of
     numerous fruit trees', *Animal Cognition* 12.6, 2009,
     pp. 797–807.

45   Lührs, Mia-Lana et al. 'Spatial memory in the grey
     mouse lemur (Microcebus murinus)', *Animal
     Cognition* 12.4, 2009, pp. 599–609.

46   Shettleworth, Sara J. 'Spatial memory in food-storing
     birds', *Philosophical Transactions of the Royal Society
     B: Biological Sciences* 329.1253, 1990, pp. 143–51.

47   Dally, Joanna M., Emery, Nathan J. and Clayton,
     Nicola S. 'Food-caching western scrub-jays keep track
     of who was watching when', *Science* 312.5780, 2006,
     pp. 1662–5.

48   Peterson, Dale. *The Moral Lives of Animals*,
     Bloomsbury Publishing USA, 2012.

49   Borgia, Gerald. 'Complex male display and female
     choice in the spotted bowerbird: specialized functions
     for different bower decorations', *Animal Behaviour*
     49.5, 1995, pp. 1291–1301.

50   Pickering, S. P. C. and Berrow, S. D. 'Courtship
     behaviour of the wandering albatross Diomedea
     exulans at Bird Island, South Georgia', *Marine
     Ornithology* 29.1, 2001, pp. 29–37.

51   Moynihan, Martin and Rodaniche, Arcadio F. 'The
     Behavior and Natural History of the Caribbean Reef
     Squid (Sepioteuthis sepioidea)', *Animal Behaviour*
     31.3, 1983, https://doi.org/10.1016/S0003-
     3472(83)80263-2.

52   Siebeck, Ulrike E. 'Communication in coral reef fish:
     the role of ultraviolet colour patterns in damselfish

territorial behaviour', *Animal Behaviour* 68.2, 2004, pp. 273–82.

53 Dixson, Danielle L., Abrego, David and Hay, Mark E. 'Chemically mediated behavior of recruiting corals and fishes: a tipping point that may limit reef recovery', *Science* 345.6199, 2014, pp. 892–7.

54 Marshall, Justin. 'Why are animals colourful? Sex and violence, seeing and signals', *Colour: Design & Creativity* 5, 2010, pp. 1–8.

55 Ghazali, Shahriman Mohd. *Fish Vocalisation: Understanding Its Biological Role from Temporal and Spatial Characteristics*, Diss, ResearchSpace, Auckland, 2011.

56 Amorim, Maria Clara C. F. Pessoa de. *Acoustic Communication in Triglids and Other Fishes*, Diss., University of Aberdeen, 1996.

57 Rowe, S. and Hutchings, Jeffrey Alexander. 'A link between sound producing musculature and mating success in Atlantic cod', *Journal of Fish Biology* 72.3, 2008, pp. 500–11.

58 Radford, Craig A. et al. 'Vocalisations of the bigeye Pempheris adspersa: characteristics, source level and active space', *Journal of Experimental Biology* 218.6, 2015, pp. 940–8.

59 Murai, Minoru, Goshima, Seiji and Henmi, Yasuhisa. 'Analysis of the mating system of the fiddler crab, Uca lactea', *Animal Behaviour* 35.5, 1987, pp. 1334–42.

60 Martinez, Francisco and Durham, Bill. 'Advantages of Reproductive Synchronization in the Caribbean Flamingo', https://socobilldurham.stanford.edu/sites/default/files/soco_-_advantages_of_reproductive_synchronization_in_the_caribbean_flamingo.pdf

61 DuVal, Emily H. 'Adaptive advantages of cooperative

courtship for subordinate male lance-tailed manakins', *American Naturalist* 169.4, 2007, pp. 423–32.

62 Martin-Wintle, Meghan S. et al. 'Free mate choice enhances conservation breeding in the endangered giant panda', *Nature Communications* 6, 2015, https://doi.org/10.1038/ncomms10125.

63 http://www.bbc.com/news/blogs-news-from-elsewhere-34733258

64 Foelix, Rainer. *Biology of Spiders*, Oxford University Press, 2010.

65 Hebets, Eileen A., Stratton, Gail E. and Miller, Gary L. 'Habitat and courtship behavior of the wolf spider Schizocosa retrorsa (Banks) (Araneae, Lycosidae)', *Journal of Arachnology*, 1996, pp. 141–7.

66 For all these examples, see Slobodchikoff, Con., *Chasing Doctor Dolittle*, op. cit., Chapter 7.

67 Darwin, Charles, Ekman, Paul and Prodger, Philip. *The Expression of the Emotions in Man and Animals*, Oxford University Press, USA, 1998.

68 Reby, David and McComb, Karen. 'Vocal communication and reproduction in deer', *Advances in the Study of Behavior* 33, 2003, pp. 231–64.

69 Reby, David et al. 'Red deer stags use formants as assessment cues during intrasexual agonistic interactions', *Proceedings of the Royal Society of London B: Biological Sciences* 272.1566, 2005, pp. 941–7.

70 Compton, L. A. et al. 'Acoustic characteristics of white-nosed coati vocalizations: a test of motivation-structural rules', *Journal of Mammalogy* 82.4, 2001, pp. 1054–8.

71 See Slobodchikoff, Con, *Chasing Doctor Dolittle*, op. cit., Chapter 2.

72 Enard, Wolfgang et al. 'Molecular evolution of
    FOXP2, a gene involved in speech and language',
    Nature 418.6900, 2002, pp. 869–72.

73 Emery, Nathan J. and Clayton, Nicola S. 'Comparing
    the complex cognition of birds and primates',
    Comparative Vertebrate Cognition, 2004, pp. 3–55.

74 Bekoff, Marc. *Minding Animals: Awareness, Emotions,
    and Heart*, Oxford University Press, 2002.

75 Hockett, Charles F. 'A system of descriptive
    phonology', *Language* 18.1, 1942, pp. 3–21.

76 Gentner, Timothy Q. et al. 'Recursive syntactic pattern
    learning by songbirds', *Nature* 440.7088, 2006,
    pp. 1204–7.

Chapter 3: Living with Animals

1   Pilley, John W. and Reid, Alliston K. 'Border collie
    comprehends object names as verbal referents',
    Behavioural Processes 86.2, 2011, pp. 184–95.

2   Pilley, John W. 'Border collie comprehends sentences
    containing a prepositional object, verb, and direct
    object', *Learning and Motivation* 44.4, 2013, pp.
    229–40.

3   Kaminski, Juliane, Call, Josep and Fischer, Julia.
    'Word learning in a domestic dog: evidence for fast
    mapping', *Science* 304.5677, 2004, pp. 1682–3.

4   For all these examples of research on dogs, see Hare,
    Brian and Woods, Vanessa, *The Genius of Dogs:
    Discovering the Unique Intelligence of Man's Best
    Friend*, Oneworld Publications, 2013.

5   Miller, Suzanne C. et al. 'An examination of changes

in oxytocin levels in men and women before and after interaction with a bonded dog', *Anthrozoös* 22.1, 2009, pp. 31–42.

6 Hearne, Vicki. *Adam's Task: Calling Animals by Name*, Skyhorse Publishing Inc., 1986.

7 Heidegger, Martin. *Zijn en tijd*, transl. Wildschut, Mark, Uitgeverij Boom, 1998.

8 Von Uexküll, Jakob. *Umwelt und Innenwelt der Tiere*, Springer-Verlag, 2014.

9 King, Barbara J. 'When animals mourn', *Scientific American* 309.1, 2013, pp. 62–7.

10 For theories of domestication see, for example, Donaldson, Sue and Kymlicka, Will, *Zoopolis: A Political Theory of Animal Rights*, Oxford University Press, 2011. For a discussion of domestication and neoteny, see Haraway, Donna Jeanne, *The Companion Species Manifesto: Dogs, People, and Significant Otherness*, Vol. 1, Chicago: Prickly Paradigm Press, 2003.

11 Donaldson, Sue and Kymlicka, Will, *Zoopolis*, op. cit.

12 Haraway, Donna Jeanne, *The Companion Species Manifesto*, op. cit.

13 Howard, Len. *Birds as Individuals*, Doubleday, 1953; Howard, Len. *Living with Birds*, Collins, 1956.

14 Lorenz, Konrad and Kerr, Marjorie. *King Solomon's Ring: New Light on Animal Ways*, Psychology Press, 2002.

15 Lorenz, Konrad, Martys, Michael and Tipler, Angelika. *Here Am I – Where Are You?: The Behavior of the Greylag Goose*, Collins, 1992.

16 Turner, Dennis C. *The Domestic Cat: The Biology of Its Behaviour*, Cambridge University Press, 2000.

17　Alger, Janet M. and Alger, Steven F. *Cat Culture: The Social World of a Cat Shelter*, Temple University Press, 2003.

18　Alger, Janet M. and Alger, Steven F. 'Beyond mead: symbolic interaction between humans and felines', *Society & Animals* 5.1, 1997, pp. 65–81.

19　Ibid.

20　See the BBC documentary *The Secret Life of the Cat* (2013) for an illustration: http://www.bbc.com/news/science-environment-22821639

21　Smith, Julie Ann. 'Beyond dominance and affection: living with rabbits in post-humanist households', *Society & Animals* 11.2, 2003, pp. 181–97.

22　Thomas, Elizabeth Marshall. *The Hidden Life of Dogs*, Houghton Mifflin Harcourt, 2010.

23　Kerasote, Ted. *Merle's Door*, Houghton Mifflin Harcourt, 2008.

24　Van Neer, Wim et al. 'Traumatism in the wild animals kept and offered at predynastic Hierakonpolis, Upper Egypt', *International Journal of Osteoarchaeology*, 2015.

25　Perry-Gal, Lee et al. 'Earliest economic exploitation of chicken outside East Asia: evidence from the Hellenistic Southern Levant', *Proceedings of the National Academy of Sciences* 112.32, 2015, pp. 9849–54.

26　Marino, Lori and Colvin, Christina M. 'Thinking Pigs: A Comparative Review of Cognition, Emotion, and Personality in Sus domesticus', *International Journal of Comparative Psychology* 28, 2015, https://escholarship.org/uc/item/8sx4s79c.

27　Smith, Carolynn L. and Johnson, Jane. 'The Chicken

Challenge: what contemporary studies of fowl mean for science and ethics', *Between the Species* 15.1, 2012, pp. 75–102.

28 Rogers, Lesley J. *The Development of Brain and Behaviour in the Chicken*, CAB International, 1995.

29 Davis, Karen. 'The social life of chickens' in *Experiencing Animal Minds: An Anthology of Animal-Human Encounters*, ed. Smith, Julie A. and Mitchell, Robert W., Columbia University Press, 2012.

30 Rogers, Lesley J. *The Development of Brain and Behaviour in the Chicken*, Wallingford, Oxfordshire, 1995, p. 48; Smith, Colin. 'Bird brain? Birds and humans have similar brain wiring', *Science Daily*, 2013, https://www.sciencedaily.com/releases/2013/07/130717095336.htm.

31 Despret, Vinciane. 'Sheep do have opinions' in Latour, B. and Weibel, P. (eds), *Making Things Public. Atmospheres of Democracy*, MIT Press, 2006, pp. 360–70.

32 Proctor, H. S. 'Measuring positive emotions in dairy cows using ear postures', http://www.researchgate.net/profile/Helen_Proctor/publication/268743762_Do_ear_postures_indicate_positive_emotional_state_in_dairy_cows/links/5475f3720cf29afed612ec7b.pdf.

33 Wathan, Jennifer and McComb, Karen. 'The eyes and ears are visual indicators of attention in domestic horses', *Current Biology* 24.15, 2014, https://doi.org/10.1016/j.cub.2014.06.023.

34 Hribal, Jason. '"Animals are part of the working class": a challenge to labor history', *Labor History* 44.4, 2003, pp. 435–53.

35 Hribal, Jason. *Fear of the Animal Planet: The Hidden History of Animal Resistance*, AK Press, 2010.

36 Hribal, Jason. 'Animals, agency, and class: writing the history of animals from below', *Human Ecology Review* 14.1, 2007, pp. 101–12.

37 Wadiwel, Dinesh. 'Do fish resist?', Human Rights and Animal Ethics Research Network, University of Melbourne, 8 December 2014.

38 See the documentary *Blackfish* for further information about Tilikum and SeaWorld.

39 Irvine, Leslie. 'The power of play', *Anthrozoös* 14.3, 2001, pp. 151–60.

40 Montaigne, Michel de. *De essays*, Singel Uitgeverijen, 2014.

Chapter 4: Thinking with the Body

1 Despret, Vinciane. 'The body we care for: figures of anthropo-zoo-genesis', *Body & Society* 10.2–3, 2004, p. 111–34.

2 Skinner, B. F. *About Behaviorism*, Vintage, 2011.

3 Chomsky, Noam. *Syntactic Structures*, Walter de Gruyter, 2002.

4 Smuts, Barbara. 'Encounters with animal minds', *Journal of Consciousness Studies* 8.5–7, 2001, pp. 293–309.

5 Candea, Matei. '"I fell in love with Carlos the meerkat": Engagement and detachment in human–animal relations', *American Ethnologist* 37.2, 2010, pp. 241–58.

6 Despret, Vinciane. 'The becomings of subjectivity in

animal worlds', *Subjectivity* 23.1, 2008, pp. 123–39.

7   Goodall, Jane. *The Chimpanzees of Gombe: Patterns of Behavior*, Belknap Press of Harvard University Press, 1986.

8   See, for example, Heinrich, Bernd, *Mind of the Raven: Investigations and Adventures with Wolf-birds*, Cliff Street Books, 1999, for love among ravens; and Würsig, Bernd, 'Leviathan love', *The Smile of a Dolphin: Remarkable Accounts of Animal Emotions*, Random House/Discovery Books, 2000, pp. 62–5, for whale love.

9   Merleau-Ponty, Maurice. *Fenomenologie van de waarneming*, transl. Tiemersma, Douwe and Vlasblom, Rens, Uitgeverij Boom, 2009.

10   Heidegger, Martin. *Zijn en tijd*, transl. Wildschut, Mark, Uitgeverij Boom, 1998.

11   Wittgenstein, Ludwig. *Filosofische onderzoekingen*, Uitgeverij Boom, 2006.

12   Hearne, Vicki. *Animal Happiness*, Perennial, 1995.

13   Martelaere, P. de. *Het dubieuze denken*, Kok/Agora, Kampen, 1996.

14   Descartes, René. *Meditaties*, Uitgeverij Boom, 1989.

15   Smith, J. David et al. 'Executive-attentional uncertainty responses by rhesus macaques (Macaca mulatta)', *Journal of Experimental Psychology: General* 142.2, 2013, p. 458.

16   Nagel, Thomas. 'What is it like to be a bat?', *Philosophical Review* 83.4, 1974, pp. 435–50.

17   Derrida, Jacques, and Mallet, Marie-Louise. *The Animal That Therefore I Am*, Fordham University Press, 2008.

18   Smuts, Barbara. 'Encounters with animal minds',
     op. cit.

Chapter 5: Structure, Grammar and Decoding

1   Mather, Jennifer A. 'Cephalopod consciousness:
     behavioural evidence', *Consciousness and Cognition*
     17.1, 2008, pp. 37–48.

2   Finn, Julian K., Tregenza, Tom and Norman, Mark D.
     'Defensive tool use in a coconut-carrying octopus',
     Current Biology 19.23, 2009, https://doi.
     org/10.1016/j.cub.2009.10.052.

3   Moynihan, Martin and Rodaniche, Arcadio F. 'The
     behavior and natural history of the Caribbean Reef
     Squid Sepioteuthis sepioidea with a consideration of
     social, signal, and defensive patterns for difficult and
     dangerous environments', *Fortschritte der
     Verhaltensforschung*, 1982.

4   Slobodchikoff, Con. *Chasing Doctor Dolittle:
     Learning the Language of Animals*, Macmillan, 2012.

5   De Saussure, Ferdinand. *Cours de Linguistique
     Générale: Edition Critique*, Vol. 1, Otto Harrassowitz
     Verlag, 1989.

6   See Slobodchikoff, Con, *Chasing Doctor Dolittle*,
     op. cit., Chapter 3. Of course, Chomsky does not
     agree with this, as he believes that language only
     occurs in humans and is not intended primarily for
     communication but to understand the world better.

7   Gentner, Timothy Q. et al. 'Recursive syntactic pattern
     learning by songbirds', *Nature* 440.7088, 2006,
     pp. 1204–7.

8   Corballis, Michael C. 'Recursion, language, and starlings', *Cognitive Science* 31.4, 2007, pp. 697–704.

9   See Slobodchikoff, Con, *Chasing Doctor Dolittle*, op. cit., pp. 197–8, 225–6.

10   Hailman, Jack P. and Ficken, Millicent S. 'Combinatorial animal communication with computable syntax: chick-a-dee calling qualifies as "language" by structural linguistics', *Animal Behaviour* 34.6, 1986, pp. 1899–1901. Also see Slobodchikoff, Con, *Chasing Doctor Dolittle*, op. cit.

11   Freeberg, Todd M., and Lucas, Jeffrey R. 'Receivers respond differently to chick-a-dee calls varying in note composition in Carolina chickadees, Poecile carolinensis', *Animal Behaviour* 63.5, 2002, pp. 837–45.

12   See Slobodchikoff, Con, *Chasing Doctor Dolittle*, op. cit., pp. 162–3.

13   Seeley, Thomas D. *Honeybee Democracy*, Princeton University Press, 2010.

14   Woo, Kevin L. and Rieucau, Guillaume. 'Aggressive signal design in the Jacky dragon (Amphibolurus muricatus): display duration affects efficiency', *Ethology* 118.2, 2012, pp. 157–68.

15   De Sá, Fábio P. et al. 'A new species of hylodes (Anura, Hylodidae) and its secretive underwater breeding behavior', *Herpetologica* 71.1, 2015, pp. 58–71.

16   Mercado III, Eduardo and Handel, Stephan. 'Understanding the structure of humpback whale songs (L)', *Journal of the Acoustical Society of America* 132.5, 2012, pp. 2947–50.

17   Suzuki, Ryuji, Buck, John R. and Tyack, Peter L. 'Information entropy of humpback whale songs',

Journal of the Acoustical Society of America 119.3, 2006, pp. 1849–66.

18 Payne, Katharine, Tyack, Peter and Payne, Roger. 'Progressive changes in the songs of humpback whales (Megaptera novaeangliae): a detailed analysis of two seasons in Hawaii', *Communication and Behavior of Whales* 10, 1987, pp. 9–57.

19 Stafford, Kathleen M. et al. 'Spitsbergen's endangered bowhead whales sing through the polar night', *Endangered Species Research* 18.2, 2012, pp. 95–103.

20 Trainer, Jill M. 'Cultural evolution in song dialects of yellow-rumped caciques in Panama', *Ethology* 80.1–4, 1989, pp. 190–204.

21 Payne, Robert B. 'Behavioral continuity and change in local song populations of village indigobirds Vidua chalybeate', *Zeitschrift für Tierpsychologie* 70.1, 1985, pp. 1–44.

22 Bohn, Kirsten M. et al. 'Versatility and stereotypy of free-tailed bat songs', *PLoS One* 4.8, 2009, https://doi.org/10.1371/journal.pone.0006746.

23 Arriaga, Gustavo, Zhou, Eric P. and Jarive, Erich D. 'Of mice, birds, and men: the mouse ultrasonic song system has some features similar to humans and song-learning birds', *PLoS One* 7.10, 2012, https://doi.org/10.1371/journal.pone.0046610.

24 Briggs, Jessica R. and Kalcounis-Rueppell, Matina C. 'Similar acoustic structure and behavioural context of vocalizations produced by male and female California mice in the wild', *Animal Behaviour* 82.6, 2011, pp. 1263–73.

25 Slobodchikoff, Con. *Chasing Doctor Dolittle*, op. cit., p. 166.

26 Neunuebel, Joshua P. et al. 'Female mice ultrasonically interact with males during courtship displays', *eLife* 4, 2015, https://doi.org/10.7554/eLife.06203.

27 Haraway, Donna Jeanne. *Primate Visions: Gender, Race, and Nature in the World of Modern Science*, Psychology Press, 1989.

28 Cooley, John R. and Marshall, David C. 'Sexual signaling in periodical cicadas, Magicicada spp. (Hemiptera: Cicadidae)', *Behaviour* 138.7, 2001, pp. 827–55.

29 Spangler, Hayward G. 'Moth hearing, defense, and communication', *Annual Review of Entomology* 33.1, 1988, pp. 59–81.

30 Von Helversen, Dagmar and Von Helversen, Otto. 'Recognition of sex in the acoustic communication of the grasshopper Chorthippus biguttulus (Orthoptera, Acrididae)', *Journal of Comparative Physiology* A180.4, 1997, pp. 373–86.

31 Huber, Franz and Thorson, John. 'Cricket auditory communication', *Scientific American* 253.6, 1985, pp. 47–54.

32 Gibson, Gabriella and Russell, Ian. 'Flying in tune: sexual recognition in mosquitoes', *Current Biology* 16.13, 2006, pp. 1311–16.

33 Kajiura, Stephen M. and Holland, Kim N. 'Electroreception in juvenile scalloped hammerhead and sandbar sharks', *Journal of Experimental Biology* 205.23, 2002, pp. 3609–21.

34 Wittgenstein, Ludwig. *Lectures and Conversations on Aesthetics, Psychology, and Religious Belief*, transl. Barrett, Cyril, University of California Press, 2007.

Chapter 6: Metacommunication

1   Bekoff, Marc. 'Social play in coyotes, wolves, and dogs', *Bioscience* 24.4, 1974, pp. 225–30.

2   Bauer, Erika B. and Smuts, Barbara B. 'Cooperation and competition during dyadic play in domestic dogs, Canis familiaris', *Animal Behaviour* 73.3, 2007, pp. 489–99.

3   Burghardt, Gordon M. *The Genesis of Animal Play: Testing the Limits*, MIT Press, 2005.

4   Massumi, Brian. *What Animals Teach Us about Politics*, Duke University Press, 2014.

5   Darwin, Charles. *The Formation of Vegetable Mould, through the Action of Worms, with Observations on their Habits*, John Murray, 1892.

6   Bekoff, Marc and Pierce, Jessica. *Wild Justice: The Moral Lives of Animals*, University of Chicago Press, 2009.

7   See Donaldson, Sue and Kymlicka, Will, 'Unruly beasts: animal citizens and the threat of tyranny', *Canadian Journal of Political Science* 47.01, 2014, pp. 23–45, for a discussion.

8   Ibid.

9   Also see Krause, Sharon R., 'Bodies in action: Corporeal agency and democratic politics', *Political Theory* 39.3, 2011, pp. 299–324.

10 See Bekoff, Marc and Pierce, Jessica, *Wild Justice*, op. cit., for a discussion of this and other examples.

11 Ibid.

12 Preston, Stephanie D. and Waal, Frans B. M. de. 'The communication of emotions and the possibility of

empathy in animals' in Post, Stephen G., Underwood, Lynn G., Schloss, Jeffrey P. and Hurlbut, William B. (eds), *Altruism and Altruistic Love*, Oxford University Press, 2002, pp. 284–308.

13 Bekoff, Marc. 'Animal emotions, wild justice and why they matter: grieving magpies, a pissy baboon, and empathic elephants', *Emotion, Space and Society* 2.2, 2009, pp. 82–5.

14 Ibid.

15 Ibid.

16 Plotnik, Joshua M. and Waal, Frans B. M. de. 'Asian elephants (Elephas maximus) reassure others in distress', *PeerJ* 2, 2014, https://doi.org/10.7717/peerj.278.

17 Peterson, Dale. *The Moral Lives of Animals*, Bloomsbury Publishing USA, 2012.

18 Park, Kyum J. et al. 'An unusual case of care-giving behavior in wild long-beaked common dolphins (Delphinus capensis) in the East Sea', *Marine Mammal Science* 29.4, 2013, https://doi.org/10.1111/mms.12012.

19 There has been no scientific research in this area, but stories can be found online, for example: http://www.dolphinsworld.com/dolphins-rescuing-humans/

20 See Bekoff, Marc and Pierce, Jessica, *Wild Justice*, op. cit., and Donaldson, Sue and Kymlicka, Will, *Zoopolis: A Political Theory of Animal Rights*, Oxford University Press, 2011.

21 See Bekoff, Marc and Pierce, Jessica, *Wild Justice*, op. cit.

22 Bshary, Redouan et al. 'Interspecific communicative

and coordinated hunting between groupers and giant
moray eels in the Red Sea', *PLoS Biol* 4.12, 2006,
https://doi.org/10.1371/journal.pbio.0040431.

23   Hart, Lynette A. and Hart, Benjamin L.
'Autogrooming and Social Grooming in Impala',
Annals of the New York Academy of Sciences 525.1,
1988, pp. 399–402.

24   Milius, Susan. 'Will groom Mom for baby cuddles',
Science News 178.12, 2010, http://dx.doi.
org/10.2307/29548936.

25   Warneken, Felix et al. 'Spontaneous altruism by
chimpanzees and young children', *PLoS Biol* 5.7,
2007, https://doi.org/10.1371/journal.pbio.0050184.

26   Warneken, Felix and Tomasello, Michael. 'Varieties of
altruism in children and chimpanzees', *Trends in
Cognitive Sciences* 13.9, 2009, pp. 397–402.

27   Bartal, Inbal Ben-Ami et al. 'Pro-social behavior in
rats is modulated by social experience', *eLife* 3, 2014,
https://doi.org/10.7554/eLife.01385.

28   Grinnell, Jon, Packer, Craig and Pusey, Anne E.
'Cooperation in male lions: kinship, reciprocity or
mutualism?', *Animal Behaviour* 49.1, 1995, pp.
95–105.

29   DeAngelo, M. J., Kish, V. M. and Kolmes, S. A.
'Altruism, selfishness, and heterocytosis in cellular
slime molds', *Ethology Ecology & Evolution* 2.4,
1990, pp. 439–43.

30   Broly, Pierre and Deneubourg, Jean-Louis.
'Behavioural contagion explains group cohesion in a
social crustacean', *PLoS Comput Biol* 11.6, 2015,
https://doi.org/10.1371/journal.pcbi.1004290.

31   Bekoff, Marc and Goodall, Jane. *The Emotional Lives*

of Animals: A Leading Scientist Explores Animal Joy, Sorrow, and Empathy – and Why They Matter, New World Library, 2008.

32   Kumlien, Ludwig. 'Reason or Instinct?', *Auk* 5.4, 1888, pp. 434–5. Kumlien discusses many examples of birds helping one another.

33   See Bekoff, Marc and Pierce, Jessica, *Wild Justice*, op. cit., for a discussion.

34   Ibid.

35   Bekoff, Marc. *Minding Animals: Awareness, Emotions, and Heart*, Oxford University Press, 2002.

36   Bateson, Melissa et al. 'Agitated honeybees exhibit pessimistic cognitive biases', *Current Biology* 21.12, 2011, pp. 1070–3.

37   Dao, James. 'After duty, dogs suffer like soldiers', *New York Times*, 1 December 2011.

38   Bradshaw, G. A. *Elephant Trauma and Recovery: From Human Violence to Liberation Ecopsychology*, ProQuest, 2005.

39   Giraffes were always thought not to make any sound, as pushing the air to make the sound up that long neck would use too much energy, but researchers recently discovered that they hum at night. See Baotic, Anton, Sicks, Florian and Stoeger, Angela S. 'Nocturnal "humming" vocalizations: adding a piece to the puzzle of giraffe vocal communication', *BMC Research Notes* 8.425, 2015, https://doi.org/10.1186/s13104-015-1394-3.

40   See King, Barbara J., *How Animals Grieve*, University of Chicago Press, 2013, for more stories and information about mourning among animals.

41   See, for example, Willett, Cynthia, 'Water and wing give wonder: trans-species cosmopolitanism', *PhaenEx*

8.2, 2013, pp. 185–208, and Schaefer, Donovan O.,
'Do animals have religion? Interdisciplinary
perspectives on religion and embodiment', *Anthrozoös*
25, sup1, 2012, https://doi.org/10.2752/175303
712X13353430377291.

42 Smuts, Barbara. 'Encounters with animal minds',
Journal of Consciousness Studies 8.5–7, 2001,
pp. 293–309.

43 Goodall, Jane. 'Primate spirituality' in Taylor, Bron
(ed.), *Encyclopedia of Religion and Nature*,
Continuum, 2005, pp. 1303–6.

44 http://www.onbeing.org/program/katy-payne-in-the-
presence-of-elephants-and-whales/transcript/7821

45 Darwin, Charles. *The Descent of Man, and Selection
in Relation to Sex*, John Murray, 1871.

46 Marino, Lori. 'Brain structure and intelligence in
cetaceans' in Brakes, Philippa, and Simmonds, Mark
Peter (eds), *Whales and Dolphins: Cognition, Culture,
Conservation and Human Perceptions*, Routledge,
2011, pp. 115–28.

47 Proctor, Darby et al. 'Chimpanzees play the ultimatum
game', *Proceedings of the National Academy of
Sciences* 110.6, 2013, pp. 2070–5.

48 Range, Friederike, Leitner, Karin and Virányi, Zsófia.
'The influence of the relationship and motivation on
inequity aversion in dogs', *Social Justice Research*
25.2, 2012, pp. 170–94.

49 Chijiiwa, Hitomi et al. 'Dogs avoid people who
behave negatively to their owner: third-party affective
evaluation', *Animal Behaviour* 106, 2015, pp. 123–7.

Chapter 7: Why We Need to Talk with the Animals

1   Donaldson, Sue and Kymlicka, Will. *Zoopolis: A Political Theory of Animal Rights*, Oxford University Press, 2011.

2   Derrida, Jacques and Mallet, Marie-Louise. *The Animal That Therefore I Am*, Fordham University Press, 2008.

3   Wolfe, Cary. *Animal Rites: American Culture, the Discourse of Species, and Posthumanist Theory*, University of Chicago Press, 2003.

4   Hobson, Kersty. 'Political animals? On animals as subjects in an enlarged political geography', *Political Geography* 26.3, 2007, pp. 250–67.

5   Seeley, Thomas D. *Honeybee Democracy*, Princeton University Press, 2010.

6   Conradt, Larissa and Roper, Timothy J. 'Group decision-making in animals', *Nature* 421.6919, 2003, pp. 155–8.

7   Ibid.

8   Bellaachia, Abdelghani and Bari, Anasse. 'Flock by leader: a novel machine learning biologically inspired clustering algorithm' in *Advances in Swarm Intelligence*, Springer, 2012, pp. 117–26.

9   Amé, Jean-Marc et al. 'Collegial decision making based on social amplification leads to optimal group formation', *Proceedings of the National Academy of Sciences* 103.15, 2006, pp. 5835–40.

10 Stueckle, Sabine and Zinner, Dietmar. 'To follow or not to follow: decision making and leadership during

the morning departure in chacma baboons', *Animal Behaviour* 75.6, 2008, pp. 1995–2004.

11 Donaldson, Sue and Kymlicka, Will. *Zoopolis*, op. cit.

12 Regan, Tom. *The Case for Animal Rights*, Springer Netherlands, 1987.

13 Nussbaum, Martha C. *Frontiers of Justice: Disability, Nationality, Species Membership*, Harvard University Press, 2009.

14 Young, Iris Marion. *Inclusion and Democracy*, Oxford University Press, 2002.

15 Young, Iris Marion. *Justice and the Politics of Difference*, University Press of Princeton, 1990.

16 I write more about this in *When Animals Speak: Toward an Interspecies Democracy*, New York University Press, 2019.

17 Yeo, Jun-Han and Neo, Harvey. 'Monkey business: human–animal conflicts in urban Singapore', *Social & Cultural Geography* 11.7, 2010, pp. 681–99.

18 Ibid., p. 14.

19 Wolch, Jennifer R. and Rowe, Stacy. 'Companions in the park', *Landscape* 31.3, 1992, pp. 16–23.

20 Holden, Steve. 'Live and learn', *Teacher*, 2010, http://works.bepress.com/steve_holden/37/.

21 Lemon, Alaina. 'MetroDogs: the heart in the machine', *Journal of the Royal Anthropological Institute* 21.3, 2015, pp. 660–79.

22 Ibid.

INDEX

From Byron, Austen and Darwin

to some of the most acclaimed and original contemporary writing, John Murray takes pride in bringing you powerful, prizewinning, absorbing and provocative books that will entertain you today and become the classics of tomorrow.

We put a lot of time and passion into what we publish and how we publish it, and we'd like to hear what you think.

Be part of John Murray – share your views with us at:

www.johnmurray.co.uk

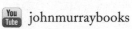 johnmurraybooks

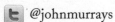 @johnmurrays

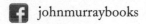 johnmurraybooks